Rawesh Kumar
Surface Characterization Techniques

Also of Interest

Polymer Surface Characterization
Sabbatini, De Giglio (Eds.), 2022
ISBN 978-3-11-070104-3, e-ISBN 978-3-11-070109-8

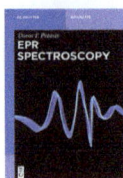

EPR Spectroscopy
Petasis, Hendrich, 2022
ISBN 978-3-11-041753-1, e-ISBN 978-3-11-041756-2

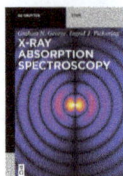

X-Ray Absorption Spectroscopy
George, Pickering, 2022
ISBN 978-3-11-057037-3, e-ISBN 978-3-11-057044-1

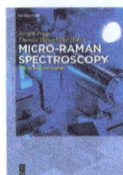

Micro-Raman Spectroscopy.
Theory and Application
Popp, Mayerhöfer (Eds.), 2020
ISBN 978-3-11-051479-7, e-ISBN 978-3-11-051531-2

Rawesh Kumar

Surface Characterization Techniques

From Theory to Research

DE GRUYTER

Authors
Dr. Rawesh Kumar
Indus University
Rancharda, Thaltej
Ahmedabad
Gujarat 382115
India

ISBN 978-3-11-065599-5
e-ISBN (PDF) 978-3-11-065648-0
e-ISBN (EPUB) 978-3-11-065658-9

Library of Congress Control Number: 2021951078

Bibliographic information published by the Deutsche Nationalbibliothek
The Deutsche Nationalbibliothek lists this publication in the Deutsche Nationalbibliografie;
detailed bibliographic data are available on the Internet at http://dnb.dnb.de.

© 2022 Walter de Gruyter GmbH, Berlin/Boston
Cover image: CasarsaGuru/E+/Getty Images
Typesetting: Integra Software Services Pvt. Ltd.
Printing and binding: CPI books GmbH, Leck

www.degruyter.com

Preface

The book *Surface Characterization Techniques: From Theory to Research* deals 10 cutting-edge surface characterization techniques with perfect academic touch. My father late Baleshwar Singh always encouraged me to write literature for everyone, and my mother Rukmani Devi stimulated me to include every detailed explanation in the literature up to a satisfactory level. That means a literature should fit with the mass as well as with class. Under that methodology, each chapter is drafted chiefly under four sections. Every chapter begins with preliminary outline of technique under section "Background" and "Instrumentation and Working Principle." Further, it extends to deep understanding under section "Glossary," and finally it ends with current research update under section "Analysis." If someone wants just peripheral knowledge of cutting-edge surface techniques, the first two sections are sufficient for them. But for academicians, research graduates and scholars, all section are equally important. That means it has contents for everyone. However, I believe that reader will decide how much extent I have succeeded to prepare this book for everyone!

In the chapters in the figure caption, I have used "right" and "left" frequently. "Right" is meant for figure at the right-hand side and "left" is meant for figure at the left-hand side. I also want to acknowledge the effort of my research students/ scholar Mayankkumar Lakshmanbhai Chaudhary and Rutu Dashrathbhai Patel in manuscript organization. This work is devoted to my believe in the Almighty Ganesh-Shiv-Shakti, my father late Baleshwar Singh and my mother Smt. Rukmani Devi.

Dr. Rawesh Kumar
Associate Professor
Indus University, Ahmedabad, Gujarat, India

https://doi.org/10.1515/9783110656480-202

Contents

1 Surface area and porosity

1.1 Background

Under high pressure and low temperature, thermal energy of gaseous molecules gets down and it is adsorbed over the entire clean surface layer-wise. The graph that represents the amount of adsorbate (adsorbed on the surface of adsorbent) on the y-axis and pressure (at a constant temperature) on the x-axis is known as an isotherm. Langmuir and Freundlich deal with the kinetics related to monolayer adsorption, that is, adsorption rate depends on the exposed area (uncovered area; $1 - \theta$), whereas desorption rate depends on the covered (θ) area on instant pressure. Brunauer–Emmett–Teller (BET) deals with kinetics related to multilayer, that is, the adsorbed molecules in the former layer act as sites of adsorption by the subsequent layer molecules. Further, the kinetics is extended to nonlimiting multilayer adsorption, and monolayer is followed by reversible-limiting-multilayer adsorption and irreversible-limiting-multilayer adsorption. As per the different absorption isotherms and the amount of adsorbate, surface area, pore volume and pore diameter of the same can be determined. The quantitative approach of "dopant loading" for covering a monolayer around a support[a] can be found by the surface area data and x-ray diffraction (XRD) pattern.

1.2 Instrumentation and working principle

Analysis sample may have contaminants (moisture and oils). So, it should be cleaned by placing the sample in a glass cell and heating it under vacuum or flowing gas. After cleaning, the sample is brought to a constant temperature by an external bath (liquid nitrogen). At low pressure, gaseous molecules have high thermal energy and high escape velocity near the surface of adsorbent. Under the condition of high pressure and low temperature, thermal energy of gaseous molecules is decreased and more number of gaseous molecules are available near the surface. Then, small amounts of a gas (the adsorbate) are admitted into the evacuated sample chamber in a stepwise manner. Gas molecules that stick to the surface of the solid (adsorbent) are said to be adsorbed and tend to form a thin layer that covers the entire adsorbent surface. After monolayer, multilayer formation (one after one layer) can be hypothesized, where previous layer is an adsorbent and the latter is the adsorbate. The working principle of adsorption on different substrates can be described in five sections.

1. Monolayer adsorption: Langmuir and Freundlich adsorption (type I)
2. Monolayer followed by nonlimiting-multilayer adsorption: BET adsorption (type II)
3. Nonlimiting multilayer adsorption: BET adsorption (type III)
4. Monolayer followed by reversible-limiting-multilayer adsorption: BET adsorption (type IV)

https://doi.org/10.1515/9783110656480-001

5. Monolayer followed by irreversible-limiting-multilayer adsorption: BET adsorption (types V and VI)

1.2.1 Monolayer adsorption/Langmuir and Freundlich adsorption (type 1 isotherm)

Langmuir states that every time adsorption and deadsorption phenomena happen on the surface. The rate of adsorption depends on how much fraction of surface is uncovered (exposed) $(1-\theta)$ and the pressure (P) of gas is applied at that instant. Similarly, the rate of deadsorption depends on how much fraction of surface is covered (θ). Again, gas adsorbed per gram of the adsorbent (x/m) is also proportional to the fraction of surface covered:

Covered Fractional Surface	Uncovered Fractional Surface	Total Surface
θ	$1-\theta$	$=1$

$$r_a = k_1(1-\theta)\,P \qquad (1.1)$$

$$r_d = k_{-1}(\theta) \qquad (1.2)$$

$$x/m = k_3(\theta) \qquad (1.3)$$

In the second way, it can be said that the rate of adsorption is equal to the difference in the rate of absorption on the ground layer and the rate of de-adsorption in the first layer. As adsorption and deadsorption are in equilibrium, both eqs. (1.1) and (1.2) can be written as follows:

$$k_1(1-\theta)\,P = k_{-1}(\theta)$$

$$\theta = \frac{k_1\,P}{k_{-1} + Pk_1} = \frac{P(k_1/k_{-1})}{1 + P(k_1/k_{-1})}; \quad k_1/k_1 - 1 = b$$

$$\theta = \frac{bP}{1 + bP}$$

From eq. (1.3), gas adsorbed per gram of adsorbent can be derived as follows:

$$\frac{x}{m} = k_3 \frac{bP}{1 + bP}$$

$$\frac{x}{m} = \frac{k_3 bP}{1 + bP}; \quad k_3 b = a$$

$$\frac{x}{m} = \frac{aP}{1 + bP}$$

For monolayer adsorption, surface coverage (θ) can be expressed by Langmuir's equation [1] as follows:

$$\theta = KP/(1 + KP)$$

where at lower pressure, $(1 + bP) \sim 1$; then, $x/m = aP$, $x/m \propto P$, which means the amount of gas adsorbed per gram is directly proportional to pressure. At high pressure, $(1 + bP) \sim bP$; $x/m = a/b$, $x/m = $ constant, which means the amount of gas adsorbed per gram is constant. So in intermediate pressure, Freundlich states that $x/m \propto P^{1/n}$ where $1/n$ has the value between 0 and 1. This is known as type I adsorption which is applicable for micropores [2]. The corresponding isotherm is known as type I isotherm (Figure 1.1). A steep uptake at low p/p^o indicates enhanced adsorbent–adsorbate interactions in micropores of molecular dimensions/ultramicropores. The flat region indicates the monolayer absorption, and the pressure at this time is called saturation pressure.

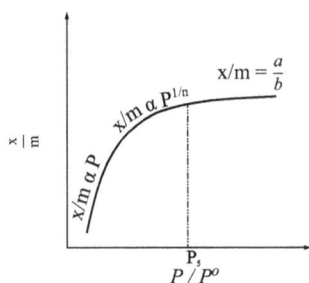

Figure 1.1: Type I isotherm.

1.2.2 Monolayer followed by nonlimiting-multilayer adsorption (type II isotherm)

The multilayer formation was explained by BET adsorption isotherm at higher pressure. The total surface (1) is composed of vacant surface sites (θ_v) as well as covered surface sites ($\Sigma_{i=1}\theta_i$).

$$1 = \theta_v + \sum_{i=1} \theta_i \qquad (1.4)$$

The covered surface sites can be quantified by the following theoretical approach:

(1) There is equilibrium between adsorption and deadsorption at the surface. Let us suppose that gas molecules (A) are adsorbed on vacant fraction sites of the ground ($S = \theta_v$); so, the first adsorbed layer coverage ($AS = \theta_1$) has been formed (Figure 1.2).

The equilibrium constant for this adsorption–deabsorption process can be derived as follows:

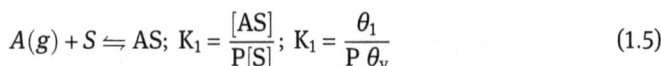

$$A(g) + S \leftrightharpoons AS; \; K_1 = \frac{[AS]}{P[S]}; \; K_1 = \frac{\theta_1}{P\,\theta_v} \qquad (1.5)$$

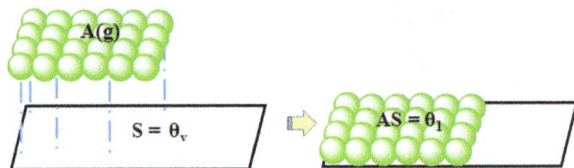

Figure 1.2: Covering of θ_1 surface by the gas molecule.

(2) For the adsorption of molecules in the higher layers, it is assumed that adsorbed molecules in the former layer act as sites for adsorption by the subsequent layer molecules (Figure 1.3). The covered surface fraction of former layer now becomes the vacant surface fraction for successive layers. Now, gas molecules (A) adsorbed on vacant sites of the first layer ($AS = \theta_1$) and second absorbed layer coverage ($A_2S = \theta_2$) have been formed. It is continued till the ith layer. Adsorption that takes place in successive layers is not much different. Hence, the equilibrium constant for successive layer absorption is equal, that is, $K_1 \cong K_2 \cdots \cong K_n \cong K$, where K is the equilibrium constant for liquefaction reaction ($A(\text{gas}) \leftrightharpoons A(\text{liquid})$; $K = 1/P^o$):

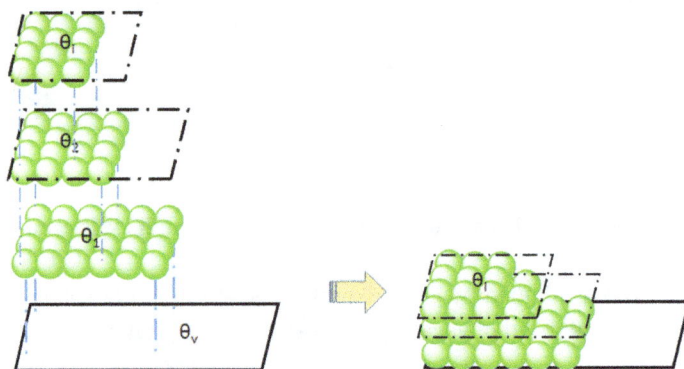

Figure 1.3: Multilayer adsorption: adsorbed molecules in the former layer act as sites for adsorption by the subsequent layer molecules.

$$A(g) + AS \leftrightharpoons A_2S; \; K_2 = \frac{[A_2S]}{P[AS]}; \; K_2 = K = \frac{\theta_2}{P\,\theta_1}; \; \theta_2 = K_2 P\,\theta_1$$

$$A(g) + A_2S \leftrightharpoons A_3S; \; K_2 = \frac{[A_3S]}{P[A_2S]}; \; K_3 = K = \frac{\theta_3}{P\,\theta_2}; \; \theta_3 = K_2 P\,\theta_2; \; \theta_3 = (KP)^2\,\theta_1$$

And so on for ith layer,

$$A(g) + A_{(i-1)}S \leftrightharpoons A_iS; \; K_2 = \frac{[A_iS]}{P[A_{(i-1)}S]}; \; K_n = \frac{\theta_i}{P\,\theta_{(i-1)}}; \; \theta_i = (KP)^{(i-1)}\,\theta_1$$

(3) Further, the covered and vacant sites can be expressed in terms of new defined constant c as $C = K_1/K$, including "c" in eq. (1.4):

$$1 = \theta_v + \sum_{i=1} \theta_i$$

$$1 = \theta_v + \sum_{i=1} (KP)^{(i-1)} \theta_1; \quad \sum_{i=1} (x)^{(i-1)} = \left(1 + x + x^2 + x^3 + \cdots\right) = 1/(1-x)$$

$$1 = \theta_v + \theta_1/(1 - KP) \tag{1.6}$$

From eq. (1.4),

$$\theta_v = \theta_1/K_1 P; \text{ (defining new constant } c = K_1/K)$$

$$\theta_v = \theta_1/cKP; \; KP = x$$

$$\theta_v = \theta_1/cx \tag{1.7}$$

Substituting eq. (1.6) into eq. (1.7)

$$1 = \theta_1/cx + \theta_1/(1-x)$$

$$\theta_1 = \frac{cx(1-x)}{1 - x + cx} = \frac{cx(1-x)}{1 + x(c-1)} \tag{1.8}$$

(4) If N_m is the number of sites (carrying the adsorbate molecule) present on per unit mass and θ_1 is the adsorbed molecule fraction, then the total number of molecules per unit mass (N) is given as follows:

$$N = N_m \sum_{i=1} i\, \theta_i = N_m\, i \sum_{i=1} (KP)^{(i-1)} \theta_1 = N_m \theta_1 \sum_{i=1} i\, (x)^{(i-1)} = N_m\, \theta_1 \left(1 + 2x + 3x^2 + \cdots\right)$$

$$\left(\text{it is known that } 1 + 2x + 3x^2 + \cdots = \frac{1}{(1-x)^2}\right)$$

$$N = N_m\, \theta_1/(1-x)^2$$

Substituting the value of θ_1 from eq. (1.8)

$$N = \left(\frac{cx(1-x)}{1 - x + cx}\right)(N_m)/(1-x)^2 = \frac{N_m cx}{(1-x)(1 + x(c-1))}$$

The number of molecules is directly proportional to the volume:

$$V = \frac{V_m cx}{(1-x)(1 + x(c-1))}$$

(Substituting $x = KP$, $K = 1/P^o$, $x = P/P^o$)

$$V = \frac{V_m cP}{(P^o - P)\left(1 + (c-1)\left(\frac{P}{P^o}\right)\right)} \tag{1.9}$$

(Note that at $P^o - P = 0$, $V \to \infty$; or infinite adsorbate volume is utilized in surface multilayer adsorption. At this place, steep rise of curve can be shown in isotherm curve):

$$\frac{V(P^o - P)}{P} = \frac{V_m c}{1 + (c - 1)(P/P^o)}$$

$$\frac{P}{V(P^o - P)} = \frac{1}{V_m c} + \frac{c - 1}{V_m c}(P/P^o)$$

The above equation can be plotted as $P/(V(P^o - P))$ versus P/P^o, where $(c - 1)/V_m c$ is the slope and $1/(V_m c)$ is the intercept of graph, respectively, as shown in Figure 1.4.

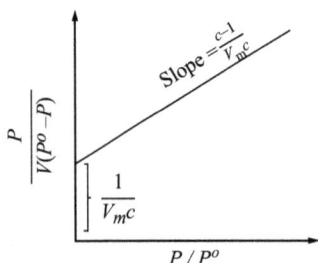

Figure 1.4: Graph of $P/(V(P^o - P))$ versus P/P^o.

(5) V_m is the volume of gas forming the unimolecular layer to cover the surface. It can be found out by $P/(V(P^o - P))$ versus P/P^o graph. At STP (standard temperature and pressure), dividing this volume by molar volume (22,400 cc) or mole of a gas molecule in a unimolecular layer can be calculated. Multiplying a mole by the Avogadro' number ($N_o = 6.023 \times 1023$), the number of gas molecules in a unimolecular layer can be found out. Finally, by multiplying this number by the cross-sectional area of N_2 gas, one can get the surface area.

Volume used in monolayer formation = V_m

Mole used in monolayer formation (at STP) = $V_m/22,400$

(on different temperatures and pressures, PV/T = constant formula can be used)

Number of gas molecules in monolayer = $\dfrac{V_m}{22,400} 6.023 \times 10^{23}$

Surface area = cross-sectional area $\times \dfrac{V_m}{22,400} 6.023 \times 10^{23}$

As the equilibrium gas pressure approaches saturation, the pores are largely/completely filled with the adsorbate. Knowing the density of the adsorbate, one can calculate the volume it occupies and, consequently, the total pore volume of the sample.

(6) V_m is the monolayer capacity and c is a constant, which determines the form of the isotherm. The parameter c is related exponentially by the equation $c = \exp [(E_1 - E_L)/RT]$.

E_1 is the heat of adsorption of gas and E_L is the heat of liquefaction of gas. The quantity $(E_1 - E_L)$ was originally known as the "net heat of adsorption" [3]. The value of "c" gives an indication of the order of magnitude for the adsorbent–adsorbate interaction. When $p/n(p^o-p)$ was plotted against p/p^o, n_m and c were obtained from the slope and the intercept in a linear BET plot.

If $E_1 > E_L$ or heat of adsorption is greater (E_1) than the heat of liquefaction (E_L), then $C \ggg 1$. It causes a rapid sharp rise of the adsorbate amount (after the flat region), indicating the multilayer formation. After the flat region, the beginning of the almost linear section corresponds to the completion of monolayer coverage. However, a gradual curvature indicates a significant amount of overlap of monolayer coverage and propagation of multilayer adsorption. At high p/p^o, the curve is characterized by nonlimiting adsorption as a result of unrestricted monolayer–multilayer adsorption. It can be understood by observing eq. (1.9):

$$V = \frac{V_m c P}{(P^o - P)\left(1 + (c-1)\left(\frac{P}{P^o}\right)\right)}$$

At $P/P^o \sim 1$, $P^o - P = 0$, $V \rightarrow \infty$, which means at high P/P^o, the infinite adsorbate volume is utilized in the surface multilayer adsorption. At this place, a steep rise of curve can be shown in the isotherm curve). This adsorption is known as type II adsorption and the corresponding isotherm is type II isotherm (Figure 1.5). This type of adsorption is found due to physisorption of gases on most nonporous or macroporous adsorbents/open surface.

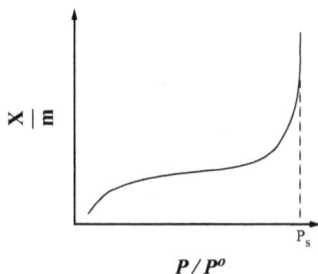

Figure 1.5: Type II isotherm.

1.2.3 Nonlimiting multilayer adsorption (type III isotherm)

If $E_L > E_1$ or heat of liquefaction (E_L) is greater than the heat of adsorption (E_1), then $C \lll 1$. It causes a rapid sharp rise of adsorbate amount (without entering in the monolayer flat region), indicating the multilayer formation. Therefore, there is no identifiable monolayer in this type of adsorption due to weak adsorbent–adsorbate interactions and the adsorbed molecules get clustered around the most favorable sites. The saturation level reaches at a pressure below the saturation vapor pressure due to prior condensation of gases in the tiny capillary pores of adsorbent at pressure below the saturation pressure (PS) of the gas. This adsorption is termed as type III adsorption and the corresponding isotherm is type III isotherm (Figure 1.6). Simply, if the flat region in the type II adsorption is missing, then the isotherm is called type III adsorption.

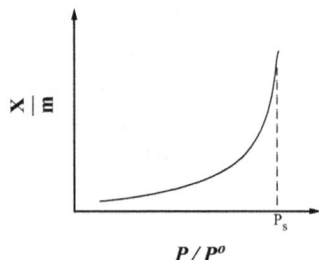

P / Pº

Figure 1.6: Type III isotherm.

1.2.4 Monolayer followed by reversible-limiting-multilayer adsorption (type IV isotherm)

A gas adsorbate may condense into a liquid-like phase in a pore at a pressure p less than the saturation pressure p^o of the bulk liquid in capillary pores of adsorbent. It leads to the appearance of a type IV adsorption and the corresponding isotherm is *type IV isotherm* (Figure 1.7). Type IV is very similar to type II except it showed limiting adsorption at high p/p^o due to the filling of mesopores by capillary condensation. It indicates the initial stages of multilayer adsorption on the mesopore walls occurred as same as over the open surface but at high p/p^o it is limited by the capillary architect of pores. This adsorption is generally characterized by hysteresis H1 and H2 loop (discussed later). Type IV isotherm indicates the large uptake of nitrogen at relative pressures between 0.5 and 0.9 p/p_0. According to the IUPAC classification, the value lies in the mesopore domain [2, 4]. A sharp inflection at a relative pressure in the range of 0.6–0.75 corresponding to the capillary condensation of N_2 indicates the uniformity of the pores of highly mesoporous material [5–7]. Again a plateau at relative pressures above 0.9 p/p_0 indicates the absence of larger mesopores, that is, macropores [4].

A reversible process of gas desorption can be carried out by withdrawing the known amounts of gas from the system in several steps. This time the traced curve is called desorption isotherms. Both adsorption–deadsorption isotherm leads to a hysteresis loop which characterizes the pore type/shapes.

Pore type can be distinguished basically into three types: (a) narrow distribution of mesopores (cylindrical pore), (b) ink-bottle pore (narrow neck exposed to the surface and wide body behind) and (c) slit-shaped pore made by nonrigid aggregates or plate-like particles. Cylindrical pore facilitates delayed capillary condensation, forming H1 hysteresis loop on isotherm. Here, desorption branch is recommended to use for pore size analysis. A material that has ink-bottle model pores shows difficulty during filtering or releasing of fluid on desorption (percolation effect). If the neck has pore size less than the critical size of adsorbent, then cavitation appears in the larger region which produces additional difficulty in fluid release from the pore during desorption (cavitation effect). These types of materials are characterized by H2 hysteresis [8, 9]. Due to some pore blocking, the desorption branch cannot be reliably used for pore size measurements. Here, adsorption branch of an isotherm is recommended to use for pore size distribution. Materials having nonrigid aggregate/slit shape/plate-like pores exhibiting unlimited absorption at high P/Po are characterized by type H3 hysteresis. H3 isotherm may also be attributed to pore blocking due to the percolation effect [10, 11]. So, again adsorption branch of isotherm is recommended to use for pore size distribution. H4 hysteresis loops are generally observed with complex materials containing both micropores and mesopores.

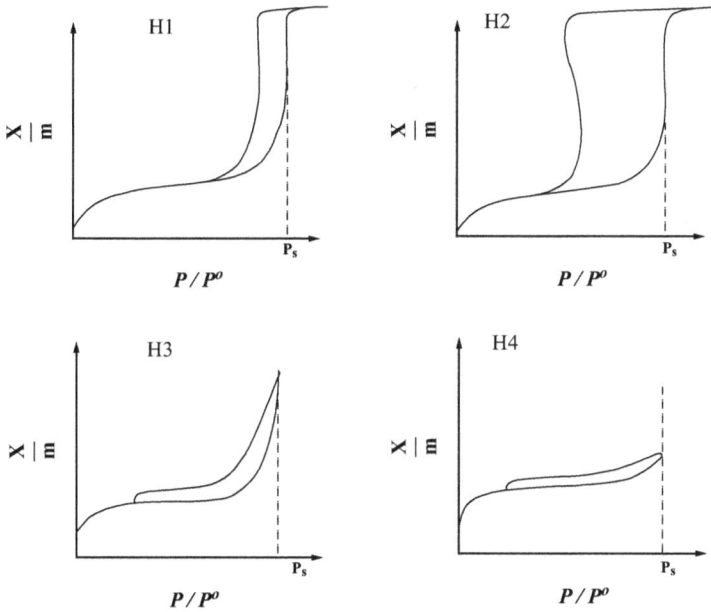

Figure 1.7: Type III isotherm having H1, H2, H3 and H4 hysteresis loops.

Interestingly, we can estimate the pore size distribution by nonlocal density functional theory model of cylindrical pore approximation [12]. For the density distribution $\rho(r)$, the excess adsorption per unit area of cylindrical pore N_s (P/P^o) can be calculated as follows:

$$N_s\left(\frac{P}{Po}\right) = \frac{2}{D_{in}}\int_0^{\frac{D}{2}} \rho(r)r\,dr - \frac{D_{in}}{4}\rho_s\left(\frac{P}{Po}\right)$$

where internal pore diameter $D_{in} = D - \sigma_{ss}$, D is the diameter between the center of oxygen atom in the external layer of pore wall and σ_{ss} is the effective diameter of oxygen. $\rho_s (P/P^o)$ is the density of bulk gas at a given relative pressure.

Further, the experimental isotherm can be represented as a combination of theoretical isotherm in individual pores as follows:

$$N_{exp} \left(\frac{P}{P^o} \right) = \int_{D_{min}}^{D_{max}} N_s \left(D_{in}, \frac{P}{P^o} \right) \varphi_s(D_{in}) dD_{in}$$

where $\varphi_s(D_{in})$ is the pore size distribution and $N_s (D_{in}, (P/P^o))$ is the kernel of theoretical equilibrium desorption isotherms in the pores of different diameters. In the same manner, second kernel of theoretical metastable adsorption isotherms was constructed. For calculating pore size distribution, these kernels were employed from the experimental desorption and adsorption isotherms, respectively. By drawing the plot between $dV(\log r)$ and "r" (where r is the radius of pore and V is the pore volume), the pore size distribution plot can be plotted. $dV(\log w)$ versus "w" (where w is the pore width) plot is also used frequently.

1.2.5 Monolayer followed by irreversible-limiting-multilayer adsorption (type V and type VI isotherm)

Type V adsorption is similar to type III except it showed limiting adsorption at high p/po due to molecular clustering after nanopore filling (mesopores), that is, adsorption shown by some activated carbon. Type VI shows the layer-by-layer adsorption, that is, Ar adsorption at low temperature at graphitized carbon black. Type V and type VI isotherms are shown in Figure 1.8.

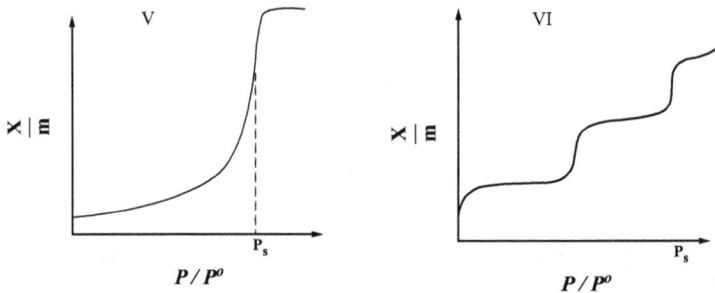

Figure 1.8: Type V and type VI isotherms.

1.3 Glossary

[a]**Monolayer coverage capacity of dopant on support (surface coverage):** The surface coverage of vanadia dopant can be calculated by the help of surface area data and XRD results. Let us take an example. Reddy et al. prepared TiO_2–ZrO_2 mixed oxide (1:1 mole ratio) supported by a homogeneous coprecipitation method using $TiCl_4$, $ZrOCl_2$, and urea [13]. Till 500 °C calcination temperature, it was amorphous as observed in XRD. Again, TiO_2–ZrO_2-supported V_2O_5 was prepared by wet impregnation method by using ammonium metavanadate over TiO_2–ZrO_2 support. It was known that 0.07 wt% per m2 vanadia is needed for monolayer coverage [14] and the TiO_2–ZrO_2 support had 160 m2/g surface area. So, 160 × 0.07 = 11.2 wt% vanadia is needed to form a monolayer. Now in XRD up to 500 °C calcination temperature, TiO_2–ZrO_2-supported V_2O_5 (<12 wt%) was found amorphous but on 12 wt% V_2O_5 load, diffraction peaks of V_2O_5 are noticed. By correlating the surface area results, V_2O_5 monolayer coverage capacity and XRD, it was concluded that V_2O_5 forms monolayer below 12 wt% loading.

1.4 Analysis of surface area and porosity results

Example 1: Palkovits et al. [15] synthesized ordered mesoporous silica (SBA-15) by following gel composition P123:tetraethylorthosilicates:H_2O:HCl = 0.16:9.54:950:5.12. On different aging, calcination and acid treatment condition, changes in the surface area and pore parameter were analyzed by N_2 sorption isotherms (Figure 1.9). Sulfuric acid-treated low-temperature-calcined (573 K) sample had additional mesopore dimensions due to higher degree of condensation achieved during the acid treatment. However, without sulfuric acid treatment, even high-temperature-calcined (823 K) samples showed lower height of isotherm as well as shift of hysteresis curve to the left which indicated the shrinkage of mesopore volume. Trimethylsilylated SBA-15 showed well-defined mesopore channel at relatively high pressure, which confirmed the absence of pore-blocking effects.

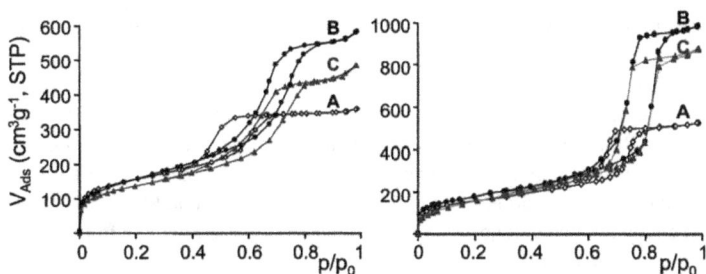

Figure 1.9: N_2 sorption isotherms of SBA-15 samples: Left (A): aged at 333 K and calcined at 823 K; left (B): aged at 333 K, H_2SO_4 treated and calcined at 573 K; left (C): aged at 333 K, H_2SO_4 treated, trimethylsilylated and calcined at 573 K; right (A): aged at 373 K and calcined at 823 K; right (B): aged at 373 K, H_2SO_4 treated and calcined at 573 K; right (C): aged at 373 K, H_2SO_4 treated, trimethylsilylated and calcined at 573 K. With permission from Elsevier.

Example 2: Sandip et al. synthesized Ga-incorporated 3D spongelike, mesoporous high-surface-area silicates (Ga-TUD-1) by sol–gel method using the following gel composition: tetraethyl orthosilicate:*m*-gallium nitrate:triethanolamine:tetraethylammonium hydroxide:H_2O = 1:0.01–0.04:3.0:0.3:15 [4]. The sample showed type IV isotherm, indicating mesopore domain (Figure 1.10). With increasing Ga loading, gallium oxide was deposited into the pores causing a decrease in pore diameter as well as pore volume.

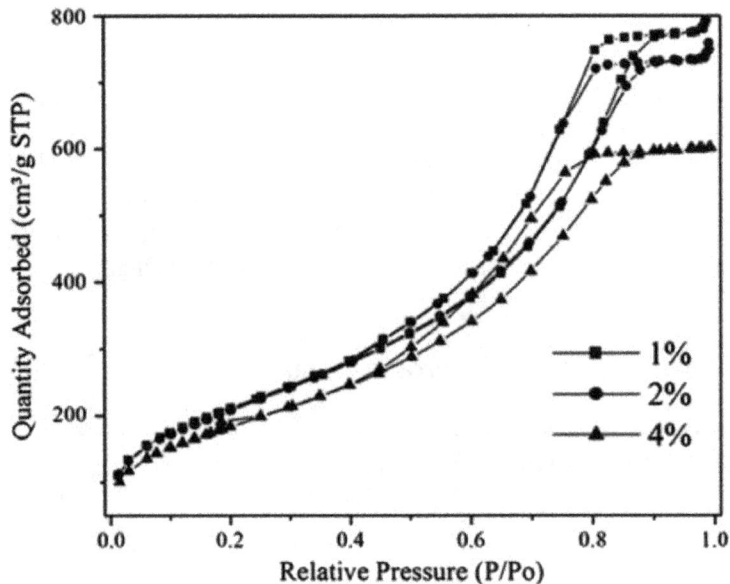

Figure 1.10: N_2 sorption isotherms of 1–4% Ga-TUD-1 samples. With permission from Elsevier.

Example 3: Rawesh et al. prepared highly ordered mesoporous niobia–zirconia composites via an evaporation-induced self-assembly method by using the following composition: $NbCl_5$:$ZrOCl_2 \cdot 8H_2O$: pluronic P123:EtOH = (0.5–4.5):(4.5–0.5):0.08:170 [16]. N_2 sorption isotherms of sample is shown in Figure 1.11. The surface area analysis showed that the mesoporous Nb_2O_5 was formed. On incorporation of 10% Zr, niobia–zirconia (Nb/Zr = 90/10) showed well-defined type IV hysteresis loop. The surface parameter (surface area, pore volume and pore diameter) was found much higher than the niobia sample. It indicated that zirconia was incorporated into niobia network and expanded the total framework. On further increase in zirconium percentage, all surface parameters fall rapidly due to the accumulation of zirconia inside and outside the pores.

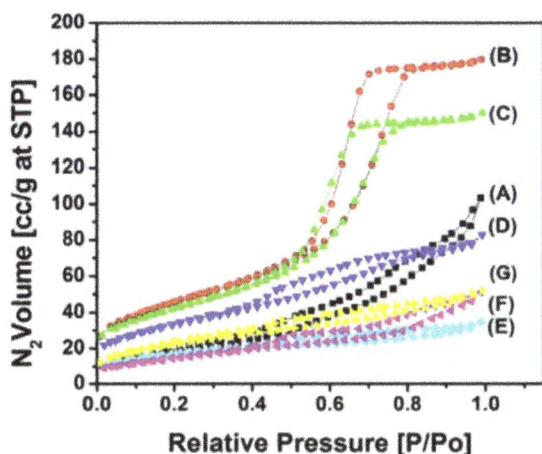

Figure 1.11: N_2 sorption isotherms of (A) Nb_2O_5, (B) Nb_2O_5–ZrO_2 (Nb/Zr = 90/10), (C) Nb_2O_5–ZrO_2 (Nb/Zr = 70/30), (D) Nb_2O_5–ZrO_2 (Nb/Zr = 50/50), (E) Nb_2O_5–ZrO_2 (Nb/Zr = 30/70) and (F) ZrO_2. With permission from Elsevier.

Example 4: Rawesh et al. prepared the high-quality hexagonally ordered mesoporous indium-containing silicates (In-SBA-15) by sol–gel method using the following gel composition: pluronic P123:H_2O:HCl: n-butanol:tetraethyl orthosilicate:In(NO$_3$)$_2$ 6H$_2$O = 0.017:200:5.4:1.325:1:0.02–0.08 [17]. N_2 sorption isotherms of sample are shown in Figure 1.12. All samples showed type IV isotherms with H1 hysteresis loop, which indicated the presence of 2D hexagonal mesoporous. Incorporation of indium up to 4 mol% in SBA-15 caused much higher surface area than only SBA-15. It indicated the successful incorporation of indium in silica matrix and expansion of silica framework thereafter. This indium loading was found to be maximum in terms of surface parameters. Upon 8 mol% indium loading, all surface parameters (surface area, pore volume and pore diameter) declined fast. Generally, concentration of cation than the optimum brought uneven folding of template by complexation with PEO-based surfactant which caused drastic fall of surface parameters.

Example 5: Mandal et al. [18] prepared manganese-doped cerium oxide by the nonhydrothermal sol–gel method using the following gel composition: triethanolamine:Ce(NO$_3$)$_3$·6H$_2$O:H$_2$O:Mn (OAc)$_2$:tetraethylammonium hydroxide = 0.2:0.1:1.1:0.006:0.1 (molar ratio) [18]. The gel was dried, calcined (at 700 °C for 10 h) and made ready for support of gold deposition by impregnation method using HAuCl$_4$·4H$_2$O at pH 7. The final material was dried and calcined (at 300 °C for 4 h) and named as 3.5 wt% Au/Mn–CeO$_2$(Mn/Ce = 2–6 mol%). N_2 sorption isotherms of sample are shown in Figure 1.13. The porosity of the catalysts is found in the 4–20 nm mesopore range. The surface area, pore volume and pore diameter of the catalyst sample had got decreased due to gold deposition onto the mesopores of Mn–CeO$_2$ support.

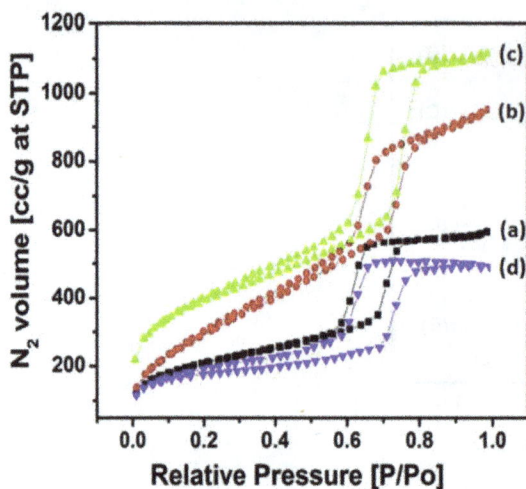

Figure 1.12: N_2 sorption isotherms of (a) SBA-15, (b) In-SBA-15 (In/Si = 2/100), (c) In-SBA-15 (In/Si = 4/100) and (d) In-SBA-15 (In/Si = 8/100). With permission from Elsevier.

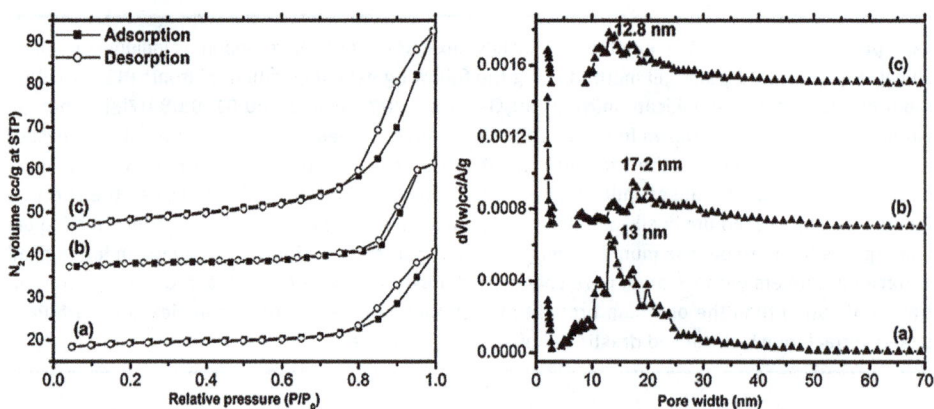

Figure 1.13: N_2 sorption isotherms and porosity of (a) Mn–CeO_2 (Mn/Ce = 2/100), (b) Mn–CeO_2 (Mn/Ce = 6/100) and (c) Au/Mn–CeO_2 (Mn/Ce = 2/100). With permission from Elsevier.

References

[1]   G. Ertl, H. Knozinger, F. Schuth, J. Weitkamp, Handbook of Heterogeneous Catalysis, WILEY-VCH, Vol. **2**.

[2]   Sing, K. S. W., Everett, D. H., Haul, R. A. W., Moscou, L., Pierotti, R. A., Rouquerol, J., Siemieniewska, T. *Handb. Heterog. Catal.* **2008**, *3–5*(4), 1503–1516.

[3]   Pechukas, P. Berichte der Bunsengesellschaft für *Phys. Chemie* **1982**, *86*(5), 372–378.

[4]   Mandal, S., Sinhamahapatra, A., Rakesh, B., Kumar, R., Panda, A., Chowdhury, B. *Catal. Commun.* **2011**, *12*(8), 734–738.

[5]   Kleitz, F., Choi, S. H., Ryoo, R. *Chem. Commun.* **2003**, *3*(17), 2136–2137.

[6]   Mandal, S., Santra, C., Kumar, R., Pramanik, M., Rahman, S., Bhaumik, A., Maity, S., Sen, D., Chowdhury, B. *RSC Adv.* **2014**, *4*(2), 845–854.

[7]   Rahman, S., Santra, C., Kumar, R., Bahadur, J., Sultana, A., Schweins, R., Sen, D., Maity, S., Mazumdar, S., Chowdhury, B. *Appl. Catal. A Gen.* **2014**, *482*, 61–68.

[8]   Santra, C., Auroux, A., Chowdhury, B. *RSC Adv.* **2016**, *6*(51), 45330–45342.

[9]   Santra, C., Pramanik, M., Bando, K. K., Maity, S., Chowdhury, B. *J. Mol. Catal. A Chem.* **2016**, *418–419*, 41–53.

[10] Ertl, G., Knözinger, H., Schüth, F., Editors, J. W., Gmbh, W. V., Kgaa, C. Techniques **2008**, *7*(4), 479–481.

[11] Santra, C., Rahman, S., Bojja, S., James, O. O., Sen, D., Maity, S., Mohanty, A. K., Mazumder, S., Chowdhury, B. *Catal. Sci. Technol.* **2013**, *3*(2), 360–370.

[12] Neimark, A. V., Ravikovitch, P. I. *Microporous Mesoporous Mater.* **2001**, *44–45*, 697–707.

[13] B. Mahipal Reddy, B. Manohar, S. M. *J. Solid State Chem.* **1992**, *97*, 233–238.

[14] N.K. NAGa, K.V.R. Chary, B. Mahipal Reddy, B. R. R. and V. S. Subrahmanyam. *Appl. Catal.* **1984**, *9* 225–233. *9*, 225–233.

[15] Palkovits, R., Yang, C. M., Olejnik, S., Schüth, F. *J. Catal.* **2006**, *243*(1), 93–98.

[16] Kumar, R., Ponnada, S., Enjamuri, N., Pandey, J. K., Chowdhury, B. *Catal. Commun.* **2016**, *77*, 42–46.

[17] Tomiyama, S., Takahashi, R., Sato, S., Sodesawa, T., Yoshida, S. *Appl. Catal. A Gen.* **2003**, *241* (1–2), 349–361.

[18] Mandal, S., Santra, C., Bando, K. K., James, O. O., Maity, S., Mehta, D., Chowdhury, B. *J. Mol. Catal. A Chem.* **2013**, *378*, 47–56.

2 Temperature-programmed surface techniques

(Reduction, hydrogenation, surface reaction, oxidation, desorption)

2.1 Background

Gases/mixture of gas blend has different interactions with sample at different temperatures. These interactions can be measured by the variation of thermal conductivity[a] of gases/gas blend before and after the interaction with sample. High thermal conductivity of gases dignifies the fast heat extract from a heated coil which drops the temperature of the coil. The temperature drop is further compensated by the current/power supplied from the external circuit (Figure 2.1A–C). Overall detector sensed a change in thermal conductivity of gases as the change in the current magnitude supplied by the external circuit for maintaining the coil at a constant temperature. The computer connected to the detector has now a set of current values vis-à-vis the temperature values (Figure 2.1D). The plot between the current (signal) and temperature is represented in this figure.

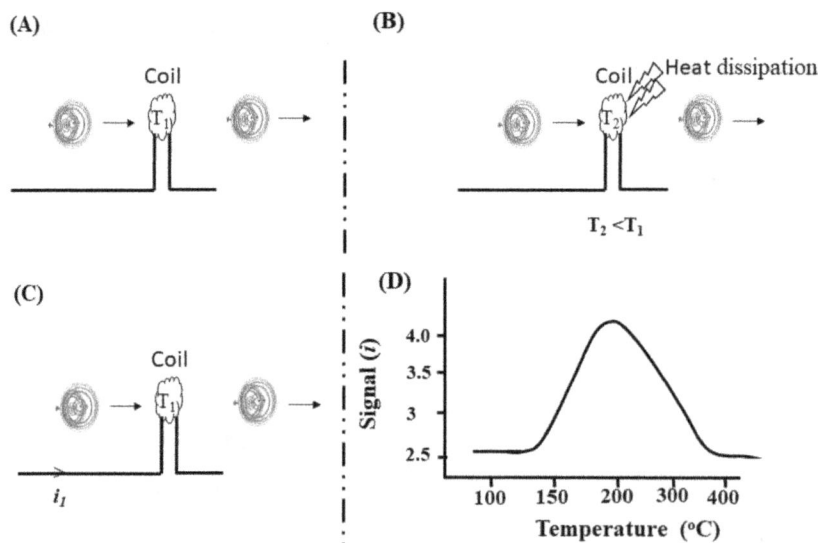

Figure 2.1: (A) Passing of gas blend through the coil maintained at T_1 temperature. (B) Heat is dissipated from the coil and temperature cools down to T_2 temperature. (C) The current is supplied from the external circuit to the coil to attain the set temperature T_1. (D) A plot between current and temperature.

https://doi.org/10.1515/9783110656480-002

2.2 Instrumentation and working principle

The schematic diagram of temperature-programmed experiment is shown in Figure 2.2. The instrument is composed of inlet line, sample cage, outlet line and heated coil connected with the external circuit. *Inlet line* receives the gases for interaction with the surface of the sample. The gas enters into the sample chamber (*sample cage*) and interacts with the sample as temperature changes. After the interaction, unadsorbed gas blend will come out through the *outlet line*, and the composition of gas blend is different in the outlet line than the inlet line. Now, the outlet gas is passed through the *heated coil*. The outlet gas removes heat (temperature drop) from the coil as per the conductivity of gas blend, that is, rapid heat liberation from coil → rapid temperature fall at coil → high conductivity of gas blend. To maintain the constant temperature of coil, the current/power is supplied by the *external circuit*. So, fall and rise of conductivity of gas blend are sensed by the detector as an increase or decrease in the amount of power supplied by the external circuit for maintaining the coil at a constant temperature. A detector also measures the difference in the thermal conductivity sensed between the gases passing over the sample and reference filaments.

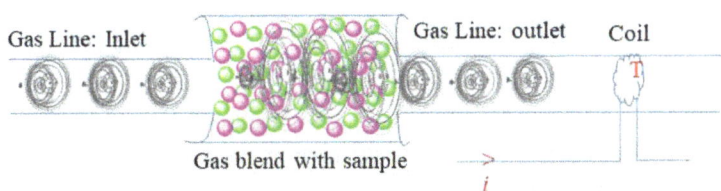

Figure 2.2: Schematic diagram of temperature-programmed experiment.

2.3 Detail procedure, working principle and data correlation

2.3.1 H$_2$-temperature-programmed reduction (H$_2$-TPR)

Temperature-programmed reduction (TPR) examines the extent to which a catalyst can be reduced or the extent to which an organic/inorganic substrate is oxidized by the catalytic surface. This analysis is carried out over a fresh catalyst.

2.3.1.1 The typical procedure
About 10% H$_2$ gas mixture in Argon gas is allowed to pass through the sample at a flow rate of 20 mL/min from ambient to 700 °C at a temperature ramp rate of 10 °C/min. Hydrogen consumption in the profiles is evaluated by comparing the peak area with calibrated files (peak area of CuO sample against different H$_2$ concentrations).

2.3.1.2 Interaction of gas blend with sample at low temperature range

At low temperature range, if gas blend is allowed to pass through the sample, there will be no adsorption of surface. The proportion of gases leaving from the sample (reaching the detector) is equal to the proportion of gas entering into the sample. Hence, conductivity of gas blend during entering and leaving (the sample) will be the same. The temperature of coil radiates with a constant rate in the gas blend environment. Here, to maintain the constant temperature of coil, constant current supply is needed. The detector records the current magnitude to maintain the constant coil temperature. Here, the current reading is constant up to some lower temperature. This plot is known as baseline reading (Figure 2.3).

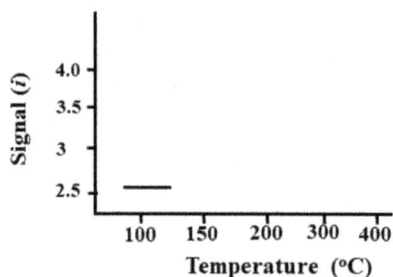

Figure 2.3: Current (signal) and temperature plot indicating baseline.

2.3.1.3 Interaction of gas blend with sample at high temperature range

Gas blend containing hydrogen can reduce the surface at high temperatures. At high temperatures, if gas blend (10% H_2 and 90% He) is allowed to pass through the sample, surface is reduced by H_2 and in turn H_2 is oxidized to H_2O (which is removed from the gas stream using cold trap). Concentration of gas blend leaving from the sample is less than the concentration of gas blend entering into the sample. It should be noted that actually leaving gas blend has less concentration of hydrogen gas. Since helium has a lower thermal conductivity than hydrogen, so the thermal conductivity of "leaving gas blend" decreases consequently.

There is a coil in the environment of leaving gas line. Temperature in the coil is brought by the current from electronic circuit outside. As conductivity of gas blend decreases, heat from the filament radiates more slowly (because heat conduction decreases). There is a need of less magnitude of current (from outside the circuit) to maintain the constant temperature. The detector records current intensity versus temperature (Figure 2.4). There is rise of *current magnitude* on maintaining higher temperature. In the graph of signal versus temperature (known as H_2-TPR profile), a steep rise of curve is shown. After a certain temperature, the interaction reaches a maximum, and then begins to diminish. Overall, a peak is generated about a temperature maximum. The instrument is already calibrated by different "H_2-adsorbed amount of concentration" versus "conductivity." By comparing, the signal intensity data can be converted into adsorbed H_2 amount (μmol/g catalyst per degree temperature) data. Overall, the adsorbed H_2 amount versus temperature graph can be plotted. This plot is known as TPR profile.

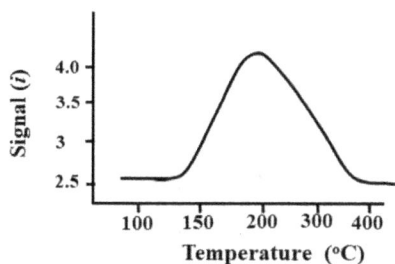

Figure 2.4: Current intensity versus temperature plot.

2.3.1.4 Chemistry during interaction of gas blend with sample

From the chemistry point of view, what is happening at the surface of sample during gas blend (H_2 containing gas) interaction at certain temperature? H_2 interacts with the surface oxygen of metal oxides. H_2 makes "bond loosely with oxygen at lower temperature" and "bond with intermediate strength with oxygen" at intermediate temperature and "bond strongly with oxygen at high temperature." Very soon, H_2 is oxidized to H_2O (which is removed from the gas stream using cold trap) and the surface is reduced (as $Ni^{+2} \rightarrow Ni^{+1} \rightarrow Ni^{\circ}$) as well as vacancies are left behind (Figure 2.5A).

As the overall hydrogen (highly conducting gas) is consumed, the conductivity of gas blend decreases after leaving from the sample. Under less conductive environment, heating coil loses its heat slowly and there is a need of small rise of current magnitude (from outside the circuit) to maintain the constant temperature. Detector detects the current intensity versus temperature. As one of the examples, Al-Fatesh et al. [1] have prepared different metal oxide solid solutions, say γ-Al_2O_3, 3 wt% TiO_2-promoted γ-Al_2O_3 and 3 wt% TiO_2-promoted γ-Al_2O_3-supported 5 wt% NiO catalyst. The H_2-TPR profile of samples is shown in Figure 2.5B. Simply, γ-Al_2O_3 and 3 wt% TiO_2-promoted γ-Al_2O_3 surface have no TPR peak, which means it is nonreducible in a wide range of temperature from –100 to 1,100 °C. In 3 wt% TiO_2-promoted γ-Al_2O_3-supported 5 wt% NiO catalyst, an intense reduction peak in the low temperature range (250–350 °C) as well as a diffuse peak in the high temperature range (400–500 °C) was observed. The lower one was attributed to the reduction of Ni^{2+} to Ni° in free NiO, whereas the higher temperature diffuse peak was attributed to the reduction of Ni^{2+} to Ni° for "NiO weakly interacted with γ-Al_2O_3 support."

2.3.1.5 Catalytic correlation

It gives the idea of self-reducibility of the catalytic surface vis-à-vis the oxidizing capability of organic/inorganic substrate by the catalytic surface. As H_2 passes from the surface, it reacts with the surface oxygen and form a H_2O molecule and left a vacancy (Figure 2.6A). Simply, it means that the surface is capable of providing surface

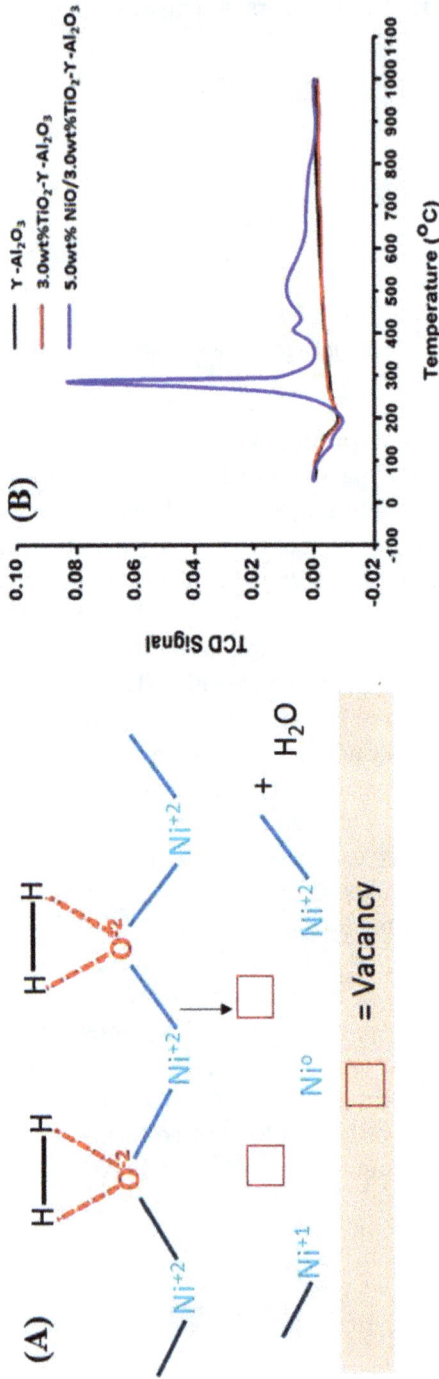

Figure 2.5: (A) H_2 interaction with the NiO surface and (B) H_2-TPR profile of γ-Al_2O_3, 3 wt% TiO_2-promoted γ-Al_2O_3 and 3 wt% TiO_2-promoted γ-Al_2O_3-supported 5 wt% Ni catalyst. With permission from Elsevier.

oxygen for a substrate or it can be said that the surface is more feasible for an oxidation reaction. When we put propylene in place of H_2, an oxidation reaction from propylene to "propylene oxide" is expected from this catalytic surface (Figure 2.6B). As observed by Chu et al. [2], CuO_x supported ordered mesoporous silicates (SBA-15) for propylene oxidation, where propylene is oxidized into propylene oxide (Figure 2.6C).

In Ni-based catalyst system, the reduced catalyst system has its importance in breaking C–H bond. As surface is more reducible, surface retains the large quantity of metallic Ni (reduced from NiO) that is responsible for CH_4 dissociation over catalyst [1] (Figure 2.6D). Such reducible surface may also oxidize the carbon deposit over the catalyst surface which is continuously formed during CH_4 dissociation at metallic nickel. If deposits are oxidized continuously, the active catalyst sites (or Nickel) are open for accepting fresh substrate CH_4 [1]. The vacancy created after reduction may be the adhering/polarizable sites for small oxidizing molecule O_2 as observed by Rahman et al. [3] over indium oxide supported on mesoporous tunable silicates (TUD-1). The vacancy created after reduction may be the adhering/polarizable sites for substrate for initializing dehydrogenation reaction as observed by Naga et al. [4] over $ZnO–CeO_2$ mixed oxide catalyst.

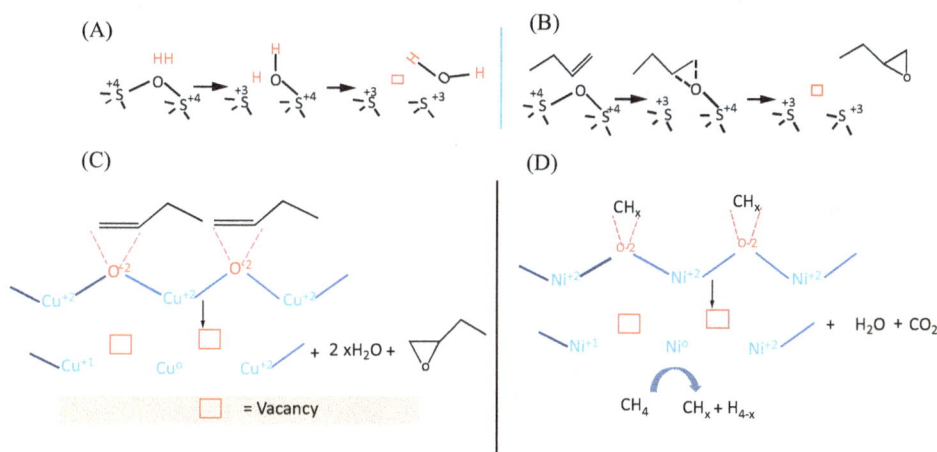

Figure 2.6: (A) Interaction of hydrogen with surface (S for surface), (B) interaction of propylene with surface, (C) interaction of propylene with surface and (D) interaction of CH_x species with surface and dissociation of CH_4 over metallic Ni.

2.3.1.6 Glossary
[a]Thermal conductivity of carrier gas blend: helium has very low thermal conductivity; hence, it is used as a carrier gas. For studying reducibility, oxidizability, acidity and basicity, specific gases H_2, O_2, NH_3 and CO_2 are used, respectively. H_2, O_2, NH_3 and CO_2 have much higher thermal conductivity. When any one of the above gas is blended with helium with fixed proportion in analysis, the gas blend has modified conductivity as per the gas mixture proportions:

$$K = \frac{K_{He}\, j_{He}\, M_{He}^{1/3} + K_{H_2}\, j_{H_2}\, M_{H_2}^{1/3}}{j_{He}\, M_{He}^{1/3} + j_{H_2}\, M_{H_2}^{1/3}}$$

where K_{He} is the thermal conductivity of helium, K_{H2} is the thermal conductivity of hydrogen, j_{He} is the mole fraction of helium, j_{H2} is the mole fraction of hydrogen, M_{He} is the molecular mass of helium and M_{H2} is the molecular mass of hydrogen.

2.3.1.7 Analysis of H$_2$-TPR profile

Example 1: Al-Fatesh et al. [1] have prepared 5 wt% Ni dispersed over "γ-alumina doped with 3 wt% SiO$_2$" by mechanically mixing of nickel nitrate precursor salt with "γ-alumina doped with 3 wt% SiO$_2$" followed by calcination. The TPR profiles of γ-Al$_2$O$_3$, SiO$_2$-modified γ-Al$_2$O$_3$ and SiO$_2$-modified γ-Al$_2$O$_3$-supported Ni catalyst samples were carried out (Figure 2.7A). Simply γ-Al$_2$O$_3$ and SiO$_2$ are nonreducible in a wide range of temperature. A peak at 300–400 °C is for the reduction of free NiO. The peak around 500–650 °C is for reduction of NiO interacted with silica. The peak around 700–850 °C is for reduction of NiO species in Al-rich phase (nickel aluminum oxide). The negative peak is due to hydrogen spillover into the mesopores of Y-Al$_2$O$_3$.

WO$_x$-modified γ-Al$_2$O$_3$-supported Ni catalysts are also prepared by the above procedure. The TPR profile of γ-Al$_2$O$_3$, WO$_3$-modified γ-Al$_2$O$_3$, WO$_3$-modified γ-Al$_2$O$_3$-supported Ni catalyst samples were carried out (Figure 2.7B). It has the absence of lower temperature peak (<400 °C) and presence of additional high-temperature peak (950 °C) with respect to silica-promoted catalyst (discussed above). The absence of low-temperature peak in WO$_x$-modified samples showed that free NiO is not present in this catalyst sample. The peak about 800 °C was previously claimed to reduction peak of NiAl$_2$O$_4$. So, at higher temperature, it was expected that some aluminum ions may be replaced and form NiWOAl-type new species. So, reduction peak about 950 °C was attributed to reduction peak of this new NiWOAl species.

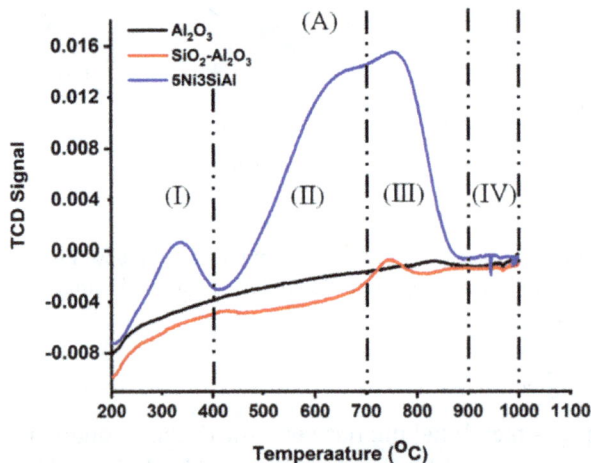

Figure 2.7: (A) H$_2$-TPR of Al$_2$O$_3$, SiO$_2$-modified γ-Al$_2$O$_3$ and SiO$_2$-modified γ-Al$_2$O$_3$-supported Ni catalyst samples and (B) H$_2$-TPR of Al$_2$O$_3$, WO$_3$-modified γ-Al$_2$O$_3$, WO$_3$-modified γ-Al$_2$O$_3$-supported Ni catalyst samples. With permission from Elsevier.

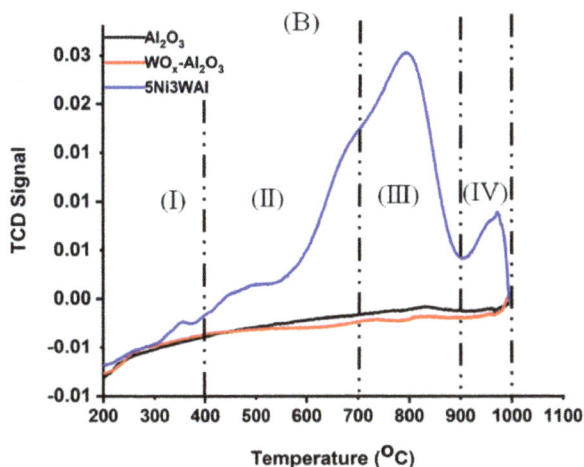

Figure 2.7 (continued)

Example 2: Al-Fatesh et al. [1] have prepared 3 wt% MoO_x-doped γ-Al_2O_3 solid solution by mechanically mixing of $(NH_4)_6Mo_7O_{24} \cdot 4H_2O$ precursor salt with γ-Al_2O_3 followed by calcination [1]. Again 5 wt% Ni (from nitrate precursor) was dispersed over MoO_x-doped γ-Al_2O_3 by the same method discussed above. The TPR profiles of γ-Al_2O_3, MoO_x-modified γ-Al_2O_3, MoO_x-modified γ-Al_2O_3-supported Ni catalyst were carried out (Figure 2.8). A peak at 300–400 °C is for the reduction of free NiO. A peak at 400–500 °C is for the reduction of Mo^{6+} to Mo^{4+}. A peak at 700–1,000 °C is for the reduction of Mo^{4+} to $Mo°$. Overlapping peaks at 500–800 °C were due to the reduction of nickel molybdenum oxide in low temperature and nickel aluminum oxide in high-temperature regions, respectively. The negative peaks are due to hydrogen spillover into the mesopores of γ-Al_2O_3.

Example 3: Shahid et al. prepared 4 mol% cobalt-doped cerium oxide by a nonhydrothermal sol–gel method using the following molar composition: triethanolamine:$Ce(NO_3)_3 \cdot 6H_2O$:H_2O:Co $(NO_3)_3 \cdot 6H_2O$:tetraethylammonium hydroxide = 0.2:0.1:1.1:(0.004–0.008–0.012):0.1 [5]. The H_2-TPR profile of CeO_2 and cobalt-doped CeO_2 were shown in Figure 2.9. The pure CeO_2 had two reduction peaks: one at low-temperature region about 673 K for reduction of surface CeO_2 and another at high temperature of 1,073 K for reduction of bulk CeO_2. In cobalt-doped ceria, low-temperature peak shifts to lower temperature, whereas high-temperature peak position remained unaffected. Apart from these, one reduction peak at 623 K for reduction of Co_3O_4 reduced to CoO species and another reduction peak at 873 K for reduction peak of species which was formed due to the interaction between CoO and CeO_2. Interestingly, a peak due to reduction of CoO to Co was not formed in this system, which was very common in the cobalt-based catalyst system. Possibly, reduction of CeO_2 to CeO_3 induced the reoxidation of Co (if formed) to CoO species in this catalyst system.

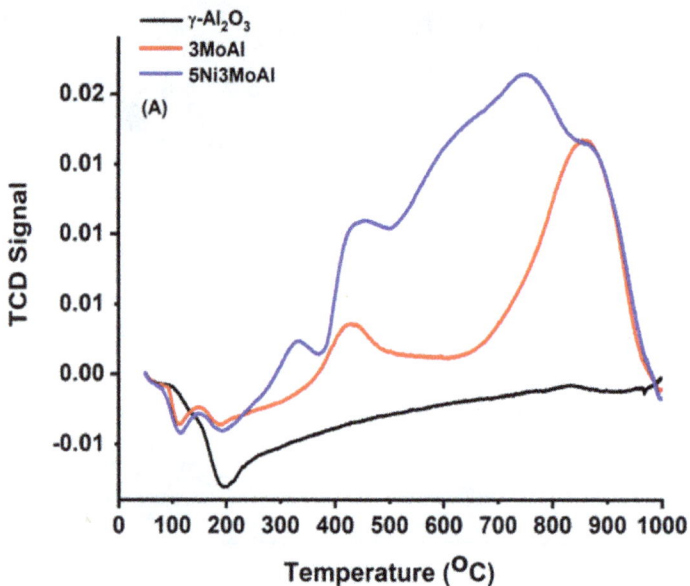

Figure 2.8: H$_2$-TPR profile of γ-Al$_2$O$_3$, MoO$_x$-modified γ-Al$_2$O$_3$ and MoO$_x$-modified γ-Al$_2$O$_3$-supported Ni catalyst. With permission from Elsevier.

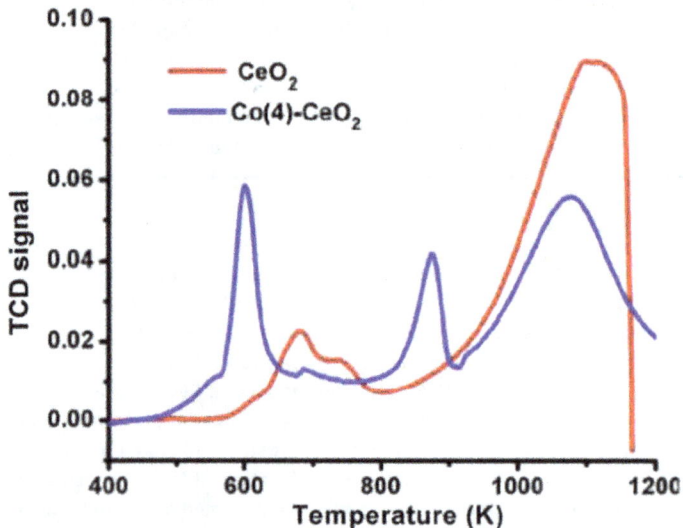

Figure 2.9: The H$_2$-TPR profile of CeO$_2$ and cobalt-doped CeO$_2$. With permission from Elsevier.

Example 4: Al-Fatesh et al. have synthesized Mg-promoted ZrO_2-supported Ni catalyst by a two-step incipient wetness impregnation. The first step is impregnation of MgO promoter over ZrO_2 support, while the second step is loading of nickel oxide over the Mg-promoted ZrO_2 support [6]. H_2-TPR profiles of ZrO_2-supported Ni and Mg-promoted ZrO_2-supported Ni are shown in Figure 2.10. ZrO_2-supported Ni showed a peak at 140–200 °C for reduction of free NiO, at 200–300 °C for NiO weakly interacted with ZrO_2 support and at 300–450 °C for NiO strongly interacted with support. After addition of MgO, a peak shift toward higher temperature indicates the formation of NiO–MgO solid solution as a peak at 450–700 °C for reduction of "NiO–MgO solid solution interacted weakly with ZrO_2 support" and at 700–900 °C for reduction of "NiO–MgO solid solution interacted strongly with ZrO_2 support." On increasing MgO loading up to 5 wt%, the later peak became prominent.

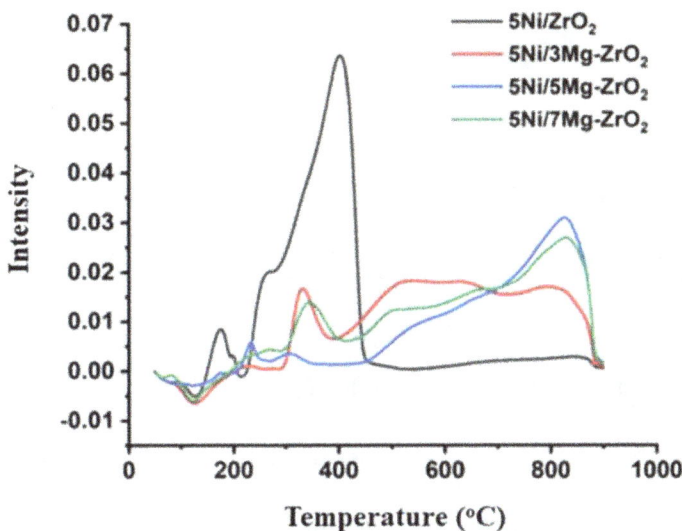

Figure 2.10: The H_2-TPR profile of ZrO_2-supported Ni, 3 wt% Mg-promoted ZrO_2-supported Ni, 5 wt% Mg-promoted ZrO_2-supported Ni and 7 wt% Mg-promoted ZrO_2-supported Ni. With permission from Nature research.

Example 5: Kasim et al. prepared ceria-promoted lanthana–zirconia-supported nickel catalyst by wet impregnation of nickel nitrate precursor and ceria nitrate precursor over lanthana–zirconia [17]. The H_2-TPR profiles of lanthana–zirconia, lanthana–zirconia-supported nickel and ceria-promoted lanthana–zirconia-supported nickel samples were shown in Figure 2.11. Ni dispersed on lanthana–zirconia support showed three NiO reduction peaks in lower (250–310 °C), intermediate (310–450 °C) and high temperature range (450–550 °C). The lower temperature peak is attributed to the reduction of "NiO species weak interaction with the support," intermediate temperature peaks for reduction of "NiO species moderately interacted with the support" and high-temperature peaks for reduction of "NiO species strongly interacted with the support." On 2.5 wt% promotional addition of ceria, reduction peaks showed a significant peak shift toward lower temperature indicating easy reducibility of all types of NiO species present on the catalyst surface.

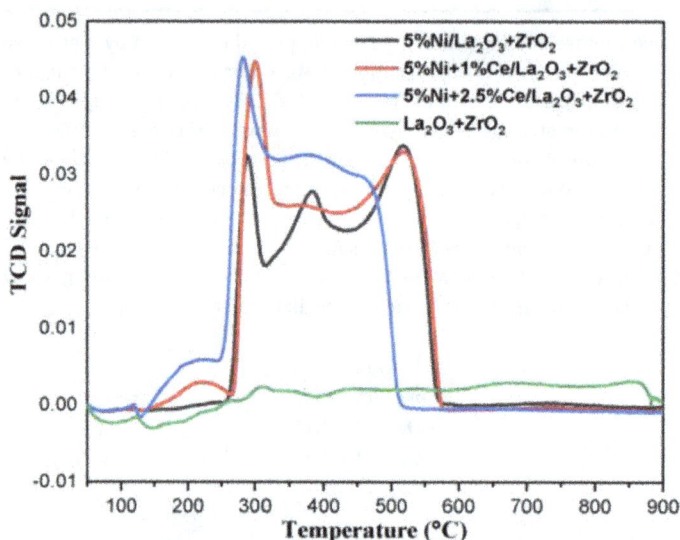

Figure 2.11: The H_2-TPR profile of lanthana–zirconia, lanthana–zirconia-supported nickel, ceria-promoted lanthana–zirconia-supported nickel samples. With permission from Elsevier.

2.3.2 Inverse temperature-programmed reduction (i-TPR)

H_2 and CO_2 interact together in methanizer chamber (having catalyst like Ni) and form methane which is easily detected by the flame ionization detector (FID) and it makes the baseline. Now, 10% H_2 in Ar is allowed to pass through the sample (which has to be analyzed). It is consumed in reducing the surface of the sample and then the remanent gas blend comes out. In outlet, it joins the CO_2 and enters into methanizer to produce CH_4 which is detected by the FID. If H_2 is consumed by the sample surface, then less amount of H_2 will come out, less amount of H_2 interacted with CO_2, less amount of CH_4 is formed and FID detects the decreased signal than the baseline. So, reducibility/consumption of H_2 by the sample is reflected by a decrease in the peak intensity (than baseline). This signal is inverse to the standard H_2-TPR (detected with TCD detector). So, this TPR surface technique is known as inverse TPR (i-TPR).

2.3.2.1 The typical procedure

About 5% H_2 gas stream is passed into the sample tube at 50 mL/min under temperature ramp from ambient to 1,000 °C. After passing the sample tube, CO_2 is mixed with the outlet stream with a rate of 2 mL/min and further the gas blend is passed through a methanizer at 573 K to produce CH_4. Further, CH_4 is detected by an FID.

2.3.2.2 Chemistry during interaction of gas blend with sample and catalytic correlation

Here, at the sample surface, H_2 interacts. So, chemistry of H_2 interaction with the surface is the same as H_2-TPR. The only difference in signal versus temperature plot is that the signal is inverse to the standard H_2-TPR (detected with TCD detector).

2.3.2.3 Analysis of i-TPR profile

Example 6: Schubert et al. prepared Pt-promoted "80 wt% Al_2O_3 and 20 wt% Co" catalyst by double-flame spray pyrolysis such that Pt remains in tight contact with Co_3O_4 [8]. Cobalt naphthenate solution (in xylene) was combusted in one flame, and aluminum-tri-sec-butoxide and Pt acetylacetonate were combusted together in the second flame. The catalyst is abbreviated as Co + xPt–Al_2O_3 ($x = 0$, 0.03, 0.08, 0.16, 0.43). The i-TPR profile of Co + xPt–Al_2O_3 ($x = 0$, 0.03, 0.08, 0.16, 0.43) is shown in Figure 2.12. Without Pt promotion, catalyst has two clear reduction peaks: about 600 K for reduction of Co_3O_4 to CoO and about 960 K for reduction of CoO to Co. On adding Pt, the corresponding reduction peaks are shifted to 525 and 695 K. With increasing Pt loading, reduction peaks were shifted to more lower temperature. It indicated the role of Pt in increasing the reducibility of catalyst surface due to high affinity of platinum for H_2 activation and a subsequent hydrogen spillover to Co_3O_4 for reduction of cobalt species.

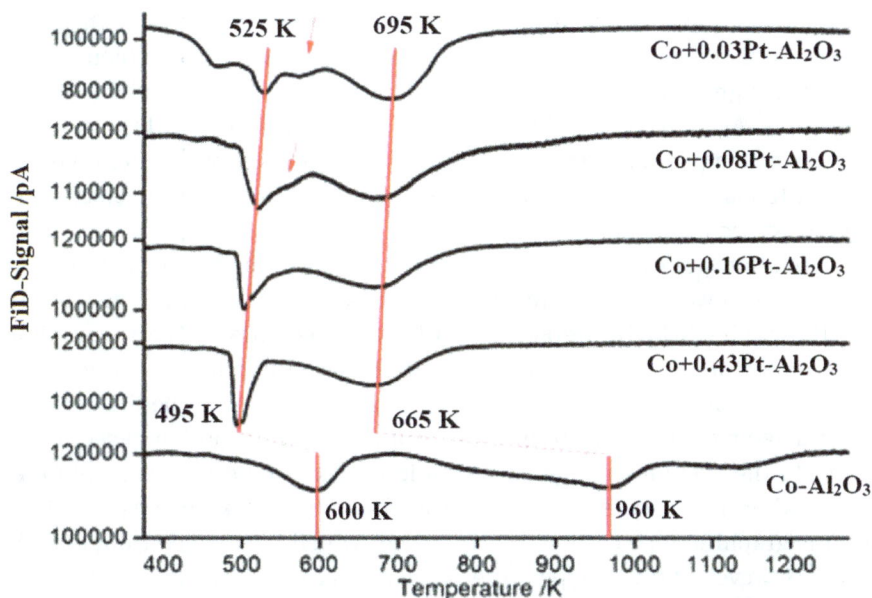

Figure 2.12: The i-TPR profile of Co + xPt–Al_2O_3 ($x = 0$, 0.03, 0.08, 0.16, 0.43). With permission from Royal Society of Chemistry.

2.3.3 Temperature-programmed hydrogenation (TPH)

Temperature-programmed hydrogenation (TPH) examines the extent of hydrogenation of carbon deposit which is created during the reaction of organic substrate over the catalyst system. It indicates the nature of carbon deposit over the spent catalyst system.

2.3.3.1 The typical procedure

The procedure is the same as H_2-TPR. The only difference is that TPH is carried out over spent catalyst system but H_2-TPR is carried out over the fresh catalyst system.

2.3.3.2 Chemistry during interaction of gas blend with sample and catalytic correlation

The outline of typical TPH procedure, surface modification over the catalyst surface and sense of change of conductivity in detector coil are presented in Figure 2.13. H_2 is passed over spent catalyst system; it may be reacted with carbon deposit and gives the hydrogenated product. If these hydrogenated products are light hydrocarbon gases (like methane, ethane, and propane), it will be evolved out from the catalyst surface, join the He stream and affect the conductivity of gas blend. The conductivity of light hydrocarbon is higher than helium; so, conductivity of gas blend increases. As more conductive gas blend passes through coil (fit at outlet), the coil loses its temperature frequently. So, extra current is supplied from the external circuit to maintain the temperature constant. The current signal versus temperature is recorded by the detector.

Hydrogenation peak at lower temperature indicates that carbon impurity is easily reducible as well as easily removable. It is known as α-carbon deposit. It may be CH_x species, which interacted lightly with the surface. Hydrogenation peak at intermediate temperature indicates that carbon impurity is moderately reducible and removable. It is known as β-carbon deposit. It may be CH_x species which interacted moderately with the surface or it may be longer but gaseous hydrocarbon. In the same manner, hydrogenation peak at high temperature indicates that carbon impurity is inert or hardly reducible. It is known as β-carbon deposit. It is graphitic-type carbon or CH_x species which interacted strongly with the surface. However, at high temperature, some metal oxide of catalyst surface was also reduced if the catalyst surface is exposed to the hydrogen stream (not covered with carbon deposit).

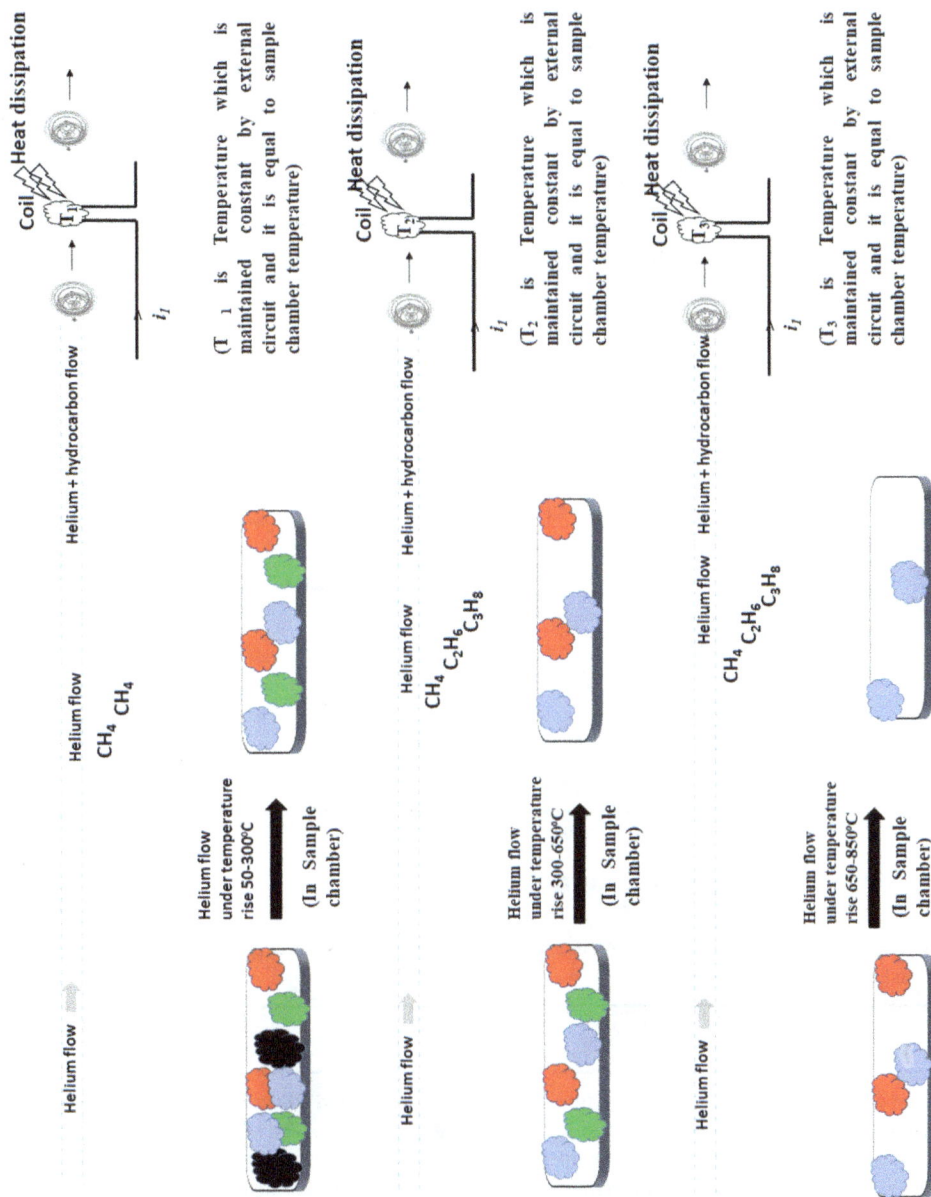

Figure 2.13: The outline of typical temperature-programmed hydrogenation procedure, surface modification over the catalyst surface and sense of change of conductivity in detector coil.

2.3.3.3 Analysis of TPH profile

Example 7: Khatri et al. prepared ceria-promoted lanthana–zirconia-supported nickel catalyst by wet impregnation of nickel nitrate precursor and ceria nitrate solutions over phosphate–zirconia [7a]. About 2.5 wt% ceria-promoted lanthana–zirconia-supported Ni catalyst was used for dry reforming of methane. The hydrogenation extent of carbon deposit over the spent catalyst system was analyzed by TPH (Figure 2.14). A sharp peak, a broad peak and a diffuse hydrogenation peak were observed about 200, 450 and 875 °C, respectively. Authors claimed that the different temperature positions of hydrogenation peaks are related to the chemical equilibrium between CH_x species ($x = 3, 2, 1$) and their activity toward hydrogen and formation of CH_4.

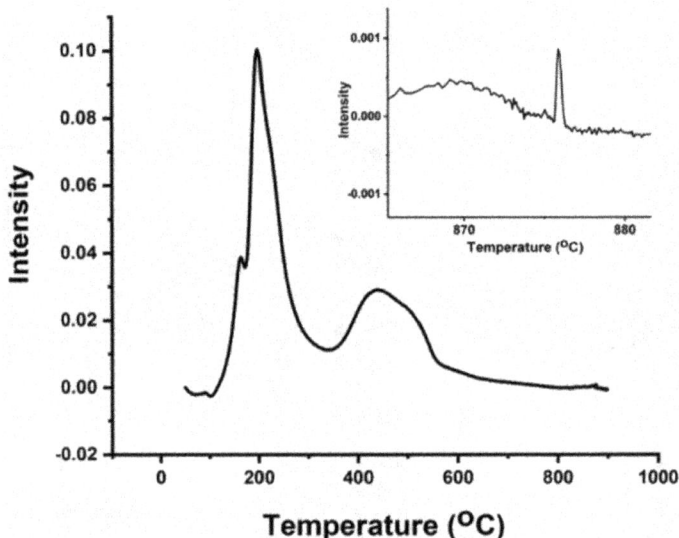

Figure 2.14: TPH profile of 2.5 wt% ceria-promoted lanthana–zirconia-supported Ni. With permission from Elsevier.

Example 8: Rutu et al. synthesized tungsten–zirconia-supported Ni catalyst and ceria-promoted tungsten–zirconia-supported Ni catalyst by mechanical crushing of metal nitrate precursor over tungsten–zirconia support followed by drying and calcination [12]. The catalysts were employed for dry reforming of methane. The hydrogenation ability of carbon deposit over the spent catalyst system was analyzed by the TPH technique (Figure 2.15). The hydrogenation peak in the range of 105–300 °C was due to hydrogenation of easily reducible and removable carbon species (called α-carbon) over the catalyst surface. Again, hydrogenation peak in the range of 390–635 °C was attributed to hydrogenation of amorphous carbon deposit (β-carbon) and hydrogenation peak about 800 °C was ascribed to hydrogenation of graphitic-type carbon (γ-carbon) or tungsten species. From the hydrogenation profile of carbon deposit, it can be said that the nonpromoted catalyst system (5 Ni/WZr) had a wide presence of moderately reducible β-carbon species, whereas 2.5 wt% Ce-promoted catalyst (5Ni2.5Ce/WZr) had prominent presence of easily reducible α-carbon species, moderately reducible β-carbon species and diffuse peaks of hardly reducible inert γ-carbon species. Clearly, ceria addition induced the reducibility of carbon deposit during a wide temperature range under the hydrogen stream range.

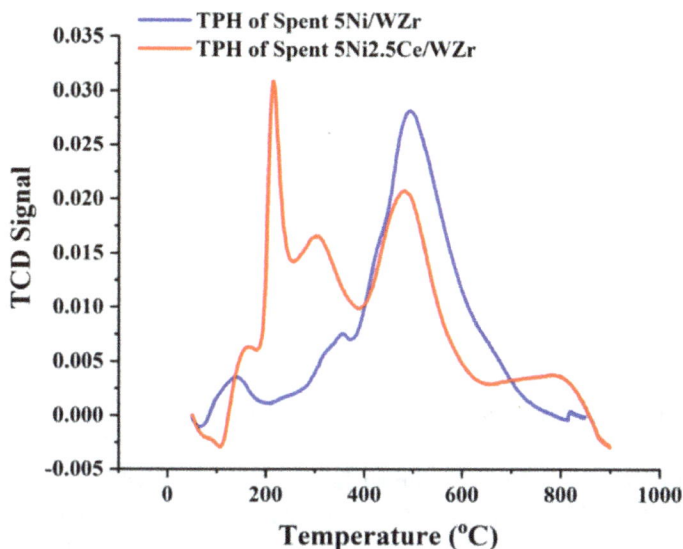

Figure 2.15: TPH profile of tungsten–zirconia-supported Ni catalyst and 2.5 wt% ceria-promoted tungsten–zirconia-supported Ni catalyst. With permission from Elsevier.

Example 9: Khatri et al. prepared ceria-promoted phosphate–zirconia-supported nickel catalyst by wet impregnation of nickel nitrate precursor and then after coimpregnation of ceria nitrate precursor over phosphate–zirconia [7b]. TPH of spent 10Ni/PZr and spent 10Ni3Ce/PZr catalyst are shown in Figure 2.16. TPH profile of spent 10Ni/PZr catalyst shows a variety of hydrogenation peak at different temperature ranges. The position of peak temperature is related to chemical equilibrium between CH_x species ($x = 3, 2, 1$) and their activity toward hydrogen. TPH of ceria-promoted catalyst has shown a greater number of reduction peaks than the unpromoted ceria catalyst. It indicates that the presence of ceria may induce more types of CH_x species (adsorbed on the surface with different strengths) to hydrogenates under H_2 stream.

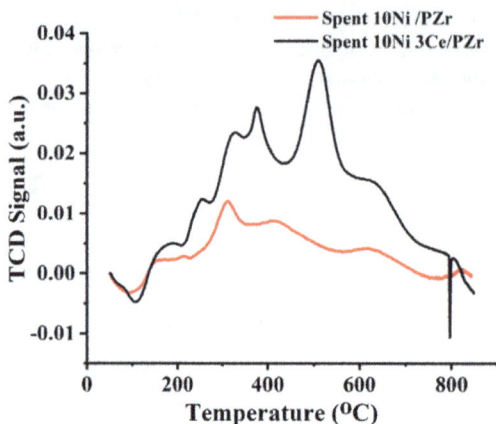

Figure 2.16: TPH profile of phosphate–zirconia-supported Ni catalyst and 2.5 wt% ceria-promoted phosphate–zirconia-supported Ni catalyst. With permission from Wiley.

2.3.4 CH$_4$-temperature-programmed surface reaction (CH$_4$-TPSR)

When CH$_4$ is allowed to blow over the catalytic surface, it interacted with the surface and at peculiar sites (may be metallic Ni, WO$_x$ species and metallic Co species), C–H bond may be broken. Overall, CH$_4$ is decomposed into hydrogen and CH$_x$ species. CH$_4$-temperature-programmed surface reaction (TPSR) gives the quantitative idea of surface sites responsible for CH$_4$ decomposition.

2.3.4.1 The typical procedure

About 10% CH$_4$ gas mixture in argon gas is allowed to pass through the sample at a flow rate of 40 mL/min from ambient to 900 °C at a temperature ramp rate 10 °C/min. CH$_4$ consumption in the profiles is evaluated by comparing the peak area with calibrated files.

2.3.4.2 Chemistry during interaction of gas blend with sample and catalytic correlation

About 10% CH$_4$ gas mixture has certain conductivity. When it reached to coil (maintained at a constant temperature by the external circuit), coil loses its temperature at a certain rate as per the conductivity of gas mixture around it. To maintain the constant temperature of coil, certain amount of current is passed through the outer circuit. This value of current makes the baseline.

When CH$_4$ gas blend is passed over sample, it interacted with the catalyst surface. At certain temperature, it is decomposed over the catalytic-active sites which is worn in the sample. It caused a decrease in the concentration of CH$_4$ in the outlet stream. So, conductivity of gas blend decreased. As less conductive gas blend passes through the coil (fit at outlet), the coil loses its temperature more slowly. To maintain the constant temperature, the magnitude of current is needed to decrease by the external circuit. Here, CH_4 consumption is indirectly reflected by the negative current signal. The current signal versus temperature is recorded by the detector. In Ni-based catalyst system, low-temperature peaks, intermediate-temperature peak and high-temperature peaks are attributed to CH_4 decomposition at metallic Ni, CH_4 decomposition at Ni−support interface and thermally driven CH_4 decomposition peak over redox support.

2.3.4.3 Analysis of CH_4-temperature-programmed surface reaction profile

Example 10: Al-Fatesh et al. synthesized Mg-promoted ZrO_2-supported Ni catalyst by a two-step incipient wetness impregnation. The first step is impregnation of MgO promoter over ZrO_2 support while the second step is loading of nickel oxide over the Mg-promoted ZrO_2 support [6]. CH_4-TPSR profile of ZrO_2, ZrO_2-supported Ni and 3 wt% Mg-promoted ZrO_2-supported Ni are shown in Figure 2.17. ZrO_2 surface showed a sharp CH_4 consumption peak at a high temperature of 870 °C. Lattice oxygen of zirconia is capable of CH_4 oxidation for a little time being at very high reaction temperature. So, this peak is attributed to thermally driven CH_4 decomposition peak over ZrO_2. ZrO_2-supported Ni catalyst had additional diffused peak at low temperature about 350 °C and a broad peak in the range of 400–800 °C and a markable peak at a high temperature of 870 °C. The low-temperature diffuse peak was claimed to the catalytic decomposition of CH_4 over Ni-active sites where broad peak in intermediate temperature was claimed to catalytic decomposition of CH_4 over Ni−Zr interface. On promotional addition of 3 wt% MgO, the CH_4-TPSR profile remained the same as unpromoted catalyst except that it had most sharp peaks at a high temperature of 870 °C. It indicates that Mg addition accelerates the thermally driven CH_4 decomposition peak over ZrO_2 also.

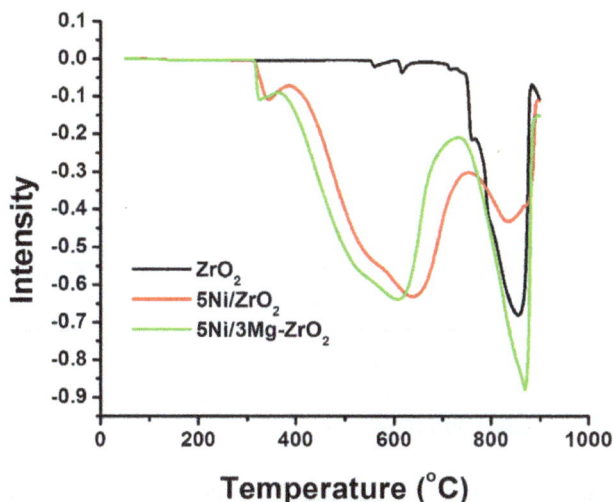

Figure 2.17: CH_4-TPSR of ZrO_2, ZrO_2-supported Ni and 3 wt% Mg-promoted ZrO_2-supported Ni. With permission from Nature research.

Example 11: Rutu et al. synthesized tungsten–zirconia-supported Ni catalyst and ceria-promoted tungsten–zirconia-supported Ni catalyst by mechanical crushing of metal nitrate precursor over tungsten–zirconia support followed by drying and calcination. CH_4-TPSR profiles of zirconia, tungsten–zirconia, tungsten–zirconia-supported Ni and 3 wt% Ce-promoted tungsten–zirconia-supported Ni are shown in Figure 2.18. ZrO_2 surface showed a sharp CH_4 consumption peak at a high temperature of 870 °C which was attributed to thermally driven CH_4 decomposition peak over ZrO_2 [12]. Tungsten–zirconia had such pronounced peak as well as an additional peak about 730 °C. Definitely, this new peak about 730 °C is due to the involvement of WO_x. So, this peak is attributed to CH_4 decomposition peak at tungsten species. Tungsten–zirconia-supported Ni catalyst showed an overlapping broad peak from low-temperature region (450 °C) to high-temperature region (850 °C). The peak at low temperature was attributed to CH_4 decomposition sites at metallic Ni, and peak at intermediate temperature range was attributed to CH_4 decomposition at an intimate contact between Ni and tungsten–zirconia (WZr) interface. The high-temperature peak was attributed to thermally driven CH_4 decomposition peak over ZrO_2. That means, tungsten–zirconia-supported Ni catalyst had a wide range of CH_4 decomposition site over the catalyst surface. On promotional addition of ceria, the CH_4-TPSR were not affected much.

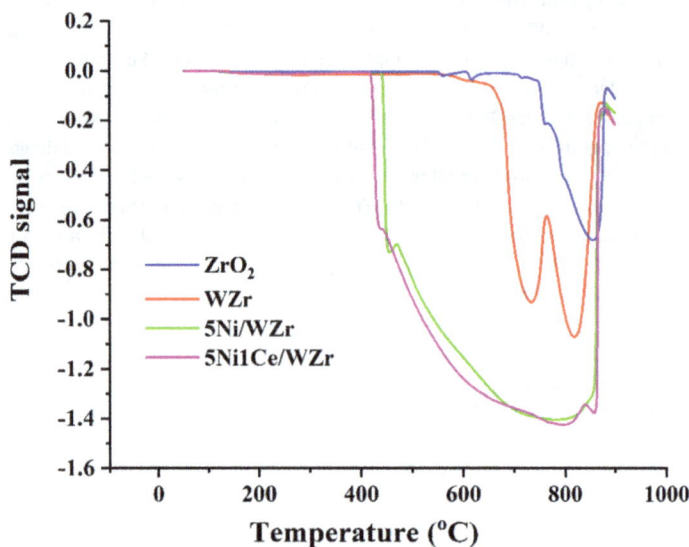

Figure 2.18: CH_4-TPSR profile of zirconia, tungsten–zirconia, tungsten–zirconia-supported Ni and 3 wt% Ce-promoted tungsten–zirconia-supported Ni. With permission from Elsevier.

2.3.5 O_2-temperature-programmed oxidation (O_2-TPO)

Temperature-programmed oxidation (TPO) examines the extent to which a catalyst surface/carbon deposit over the catalyst can be oxidized.

2.3.5.1 The typical procedure

A typical 10% O_2 gas mixture in He gas is allowed to pass through the sample (fresh/ spent) at a flow rate of 20 mL/min flow rate from ambient to 700 °C at a temperature ramp rate 10 °C/min. Oxygen consumption in the profiles is evaluated by comparing the peak area with calibrated files (peak area of Cu_2O sample against different O_2 concentrations).

2.3.5.2 Interaction of gas blend with sample at low temperature range

Gas blend containing oxygen can oxidize the surface/organic deposit over surface at high temperature. At high temperature, if gas blend (10% O_2 and 90% He) is allowed to pass through the sample, surface/organic deposit is oxidized by O_2. The metal is oxidized into metal oxide, and metal oxide is oxidized into its higher oxidation states and carbon deposit is oxidized into CO or CO_2. Overall, *oxygen (which is more conductive than argon) is consumed* by the sample/organic deposit. Proportion of oxygen gases leaving from the sample *is less* than the proportion of O_2 gas entering into the sample; so, the conductivity of gas blend decreases after leaving from the sample.

There is a coil in the environment of leaving gas line. Temperature in the coil is brought by the current from the electronic circuit outside. As the *conductivity of gas blend decreases*, heat from the filament radiates more slowly. There is a need of less magnitude of current (from outside the circuit) to maintain the constant temperature. The detector records current intensity versus temperature (Figure 2.19). There is an increase in the *current magnitude* on maintaining higher temperature. In the graph of signal versus temperature (known as O_2-TPO profile), a steep rise of curve is shown. After a certain temperature, the interaction reaches a maximum, then begins to diminish. Overall, a peak is generated about a temperature maximum. The instrument is already calibrated by various "O_2-adsorbed amount concentrations" versus "conductivities." By comparing, the signal intensity data can be converted into adsorbed H_2 amount (μmol/g catalyst per degree temperature) data. Overall, the adsorbed H_2 amount versus the temperature graph can be plotted.

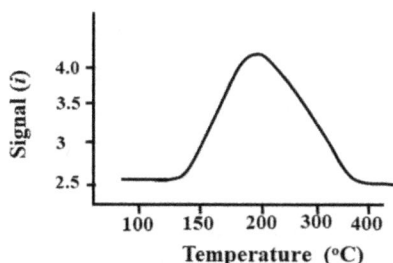

Figure 2.19: Current intensity versus temperature plot.

2.3.5.3 Chemistry during interaction of gas blend with sample and catalytic correlation

From chemistry point of view, what is happening at the surface of sample during gas blend (O_2 containing gas) interaction at certain temperature? As O_2 interacts with the prereduced surface (by H_2-TPR) at certain temperature, prereduced surface oxidation takes place with O_2. O_2 interacts with the surface vacancy of prereduced metal oxides and fills the vacancy quickly. Very soon the prereduced surface (during TPR) is oxidized again $Cu^{\circ} \rightarrow Cu^{1+} \rightarrow Cu^{2+}$ (Figure 2.20(1)). As the overall highly conducting O_2 gas is consumed, the conductivity of gas blend decreases after leaving from the sample. In the decreased conductible environment, heating coil loses its heat slowly and there is a need of small rise of current magnitude to maintain the constant temperature. The detector detects current intensity versus temperature. Consumption of H_2 mole amount in TPR vis-à-vis O_2 mole amount in TPO indicates redox potential and reversibility of oxidation–reduction cycle. As one of the examples, Rahman et al. [9] have prepared indium containing mesoporous tunable silicates (In-TUD-1). At higher indium loading, O_2 consumption per gram of catalyst dropped steeply and the oxygen chemisorption peak also shifts to higher temperature. The TPO observation indicates the easy accessibility of active sites at lower indium loading than higher loading.

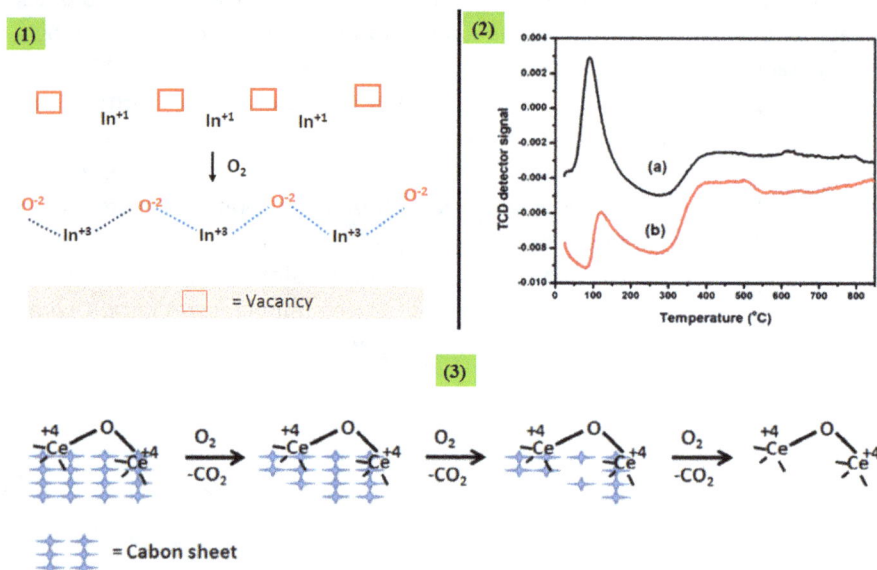

Figure 2.20: (1) Oxygen uptake by the reduced catalyst having oxygen vacancy. (2) (a) Temperature-programmed oxidation of TUD-1-supported indium catalyst having In/Si = 1/100 and (b) temperature-programmed oxidation of TUD-1-supported indium catalyst having In/Si = 4/100. TUD-1 is a mesoporous silicate matrix with a large surface area and high thermal stability. (3) Oxygen uptake by the "deposited carbon on catalyst surface" or "carbon-supported catalyst system". With permission from Elsevier.

TPO is also a useful technique to find the information regarding the nature of carbon deposit at the catalyst. Then in place of O_2 consumption, CO or CO_2 evolution per gram per degree temperature can also be monitored (Figure 2.20(2)). The amount of deposited carbon in the spent catalyst and quantification of carbon in the carbon-supported catalyst were confirmed by TPO analysis by calculating an equivalent amount of CO_2 evolution due to oxidation of carbon sheet by O_2 (used in TPO) (Figure 2.20(3)). Minh et al. [10] carried out TPO of a cyclohexene-coked cracking catalyst and quantified the evolution of CO and CO_2.

2.3.5.4 Analysis of temperature-programmed oxidation profile

Example 12: Negative peak in TPO: Anis H. Fakeeha et al. synthesized metal (M = Cu, Ga, Gd, Zn)-promoted Al_2O_3-supported Ni catalyst by dry impregnation using metal nitrate solutions with meso-γ-Al_2O_3 [11]. The TPO profile of the spent catalyst had positive and negative peaks. A typical TPO profile is shown in Figure 2.21. The negative peaks around 400 °C indicated oxidation of metals [11], whereas positive peaks below 250 °C were attributed to less ordered easily oxidizable atomic carbon.

Example 13: Rutu et al. had synthesized tungsten–zirconia-supported nickel catalyst (5Ni/WZr) and 2.5 wt% ceria-promoted tungsten–zirconia-supported nickel catalyst (5Ni2.5Ce/WZr) by the wet impregnation method using metal nitrate precursor over tungsten–zirconia support [12]. TPO profiles of the spent 5Ni/WZr and 5Ni2.5Ce/WZr materials are shown in Figure 2.22. O_2-TPO of spent 5Ni/WZr and 5Ni2.5Ce/WZr showed the peak in the range of 530–575 °C which is attributed to moderately oxidizable carbon species (β-carbon species). Spent ceria-promoted catalyst (5Ni2.5Ce/WZr) showed less intensity peak than the catalyst without ceria (5Ni/WZr). It indicated that ceria-promoted catalyst has less amount of carbon deposit because carbon deposit is regularly oxidized by the mobile oxygen endowed by ceria.

Example 14: Rutu et al. had prepared support by mixing x wt% yttria and (100–x) wt% zirconia (where x = 0, 5, 10, 15, 20 wt%) together [13]. Different amounts of yttria were used in the support mixing. Further, 5 wt% Ni (from nitrate precursor) was impregnated over support. O_2-TPO of yttria–zirconia-supported Ni samples were carried out by 10% O_2 in Ar. The O_2-TPO of yttria–zirconia-supported Ni samples are shown in Figure 2.23. O_2-TPO profile showed a major peak at 550 °C for moderately oxidizable β-carbon species. As yttrium portion is increased in ZrO_2 as support, lower intensity peaks are observed. It indicates the presence of lower amount of carbon deposit over the catalyst surface at higher yttria presence in the support. That means over Y containing system, carbon deposit formed during the reaction is oxidized by CO_2 as well as with lattice oxygen of yttria–zirconia support.

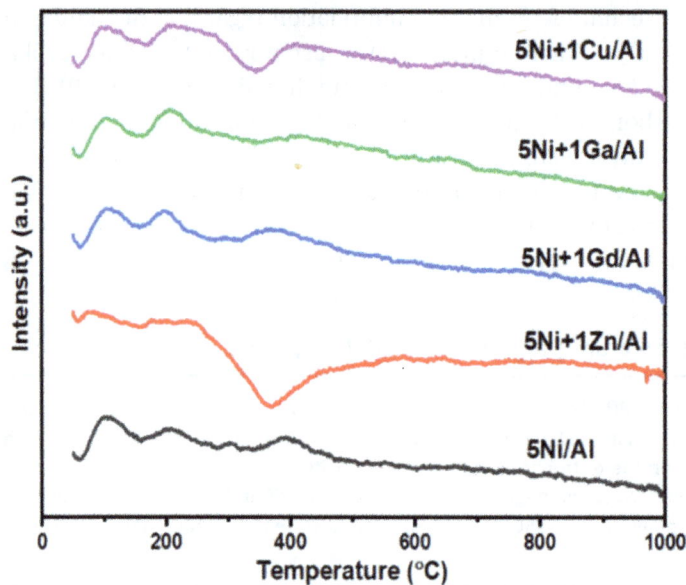

Figure 2.21: TPO profile is Al_2O_3-supported Ni catalyst, Zn-promoted Al_2O_3-supported Ni catalyst, Gd-promoted Al_2O_3-supported Ni catalyst, Ga-promoted Al_2O_3-supported Ni catalyst and Cu-promoted Al_2O_3-supported Ni catalyst. With permission from MDPI.

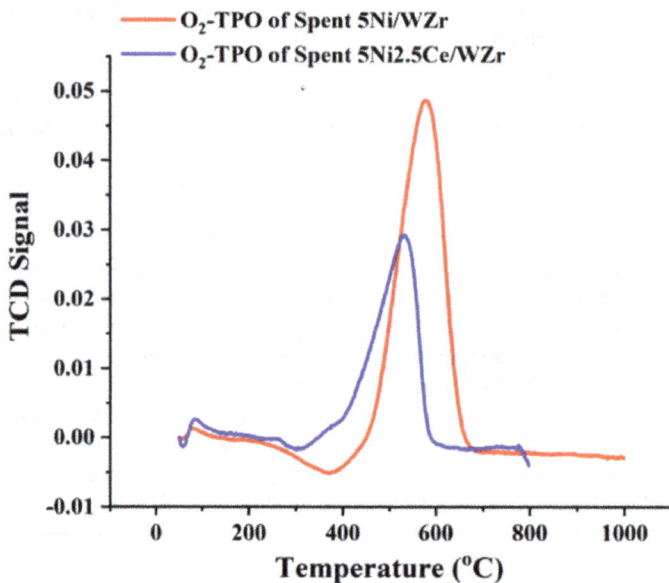

Figure 2.22: TPO profile of spent 5Ni/WZr and 5Ni2.5Ce/WZr materials. With permission from Elsevier.

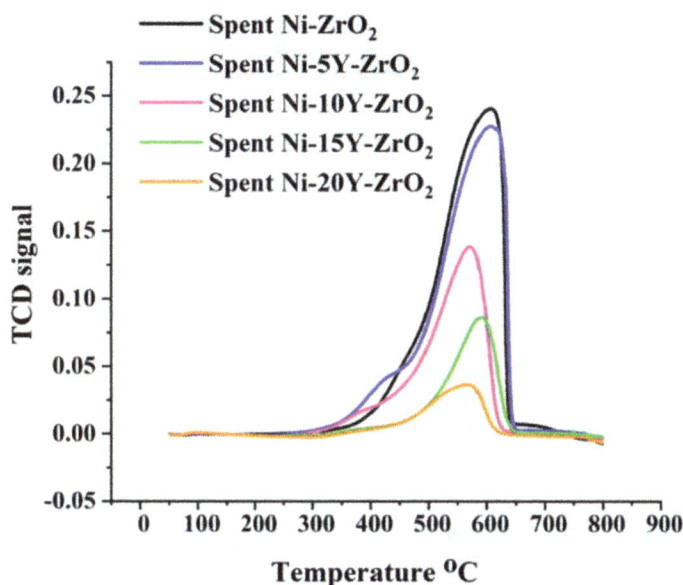

Figure 2.23: The O_2-TPO of yttria–zirconia-supported Ni samples. With permission from Elsevier.

Example 15: Khatri et al. prepared ceria-promoted lanthana–zirconia-supported nickel catalyst by wet impregnation of nickel nitrate precursor and then after coimpregnation of ceria nitrate precursor over lanthana–zirconia [7a]. It will be interesting to view the effect of presence of H_2 stream and CO_2 stream over the spent catalyst system during the dry reforming reaction. The O_2-TPO profile of spent 2.5 wt% ceria-promoted lanthana–zirconia-supported Ni sample (spent 5Ni2.5Ce/LaZr) is shown in Figure 2.24. If simply, O_2-TPO of spent ceria-promoted phosphate–zirconia-supported nickel catalyst was carried out, one peak at 400 °C due to easily oxidizable amorphous carbon species (intermediate of methane decomposition species) and another peak at about 500 °C due to moderately oxidizable carbon species (less inert carbon species) are formed. Interestingly, if O_2-TPO is carried out after CO_2-TPD, higher intense respective peaks were found. It indicated that during interaction of CO_2 with carbon deposit, higher molecular weight oxygenated carbon species were formed, which was oxidized later during O_2-TPO. Again, if O_2-TPO is carried out after TPH, most intense respective peaks were found. It indicated that during interaction of H_2 with carbon deposit, higher molecular weight hydrogenated carbon species which are strongly interacted with surface were formed which was oxidized later during O_2-TPO.

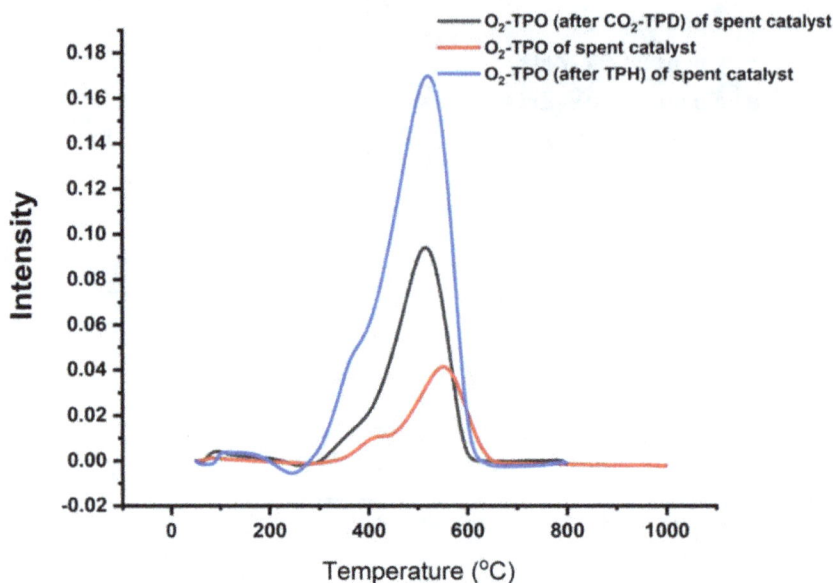

Figure 2.24: O_2-TPO profile of spent 5Ni2.5Ce/LaZr in different conditions. With permission from Elsevier.

2.3.6 NH$_3$-temperature programmed de-adsorption (NH$_3$-TPD)

To illustrate the acidic property of the surface, basic probes such as NH_3 and C_5H_5N (pyridine) are used under temperature-programmed de-adsorption (TPD) techniques. NH_3 is basic in nature which can be interacted with an acidic surface. It adsorbed weakly with weak acid sites, moderately with intermediate-strength acid sites and strongly with strong acid sites (Figure 2.25). If surface acidity is expected to wear by hydroxyl, then surface hydroxyl contributes weak acidity, bridging hydroxyl contributes intermediate-strength acidity and metal-coordinated hydroxyl contributes strong acid sites. With increasing temperature, NH_3 would be de-adsorbed from the surface as per its interaction extent with the surface. Temperature-programmed desorption of NH_3 examines the acidic strength of surface at different temperatures.

2.3.6.1 The typical procedure

The sample was first degassed in flow of helium for 2 h at 200 °C. Now, 10% NH_3 in He is allowed to pass through the sample at room temperature for 30 min. Afterward, the excess NH_3 was flushed out in a flow of He for 45 min. Then the temperature-programmed desorption of ammonia (carrier gas He) was obtained by heating from ambient temperature to 1,000 °C at a temperature ramp of 10 °C/min. The thermal conductivity of gas blend is detected by the thermal conductivity detector (TCD).

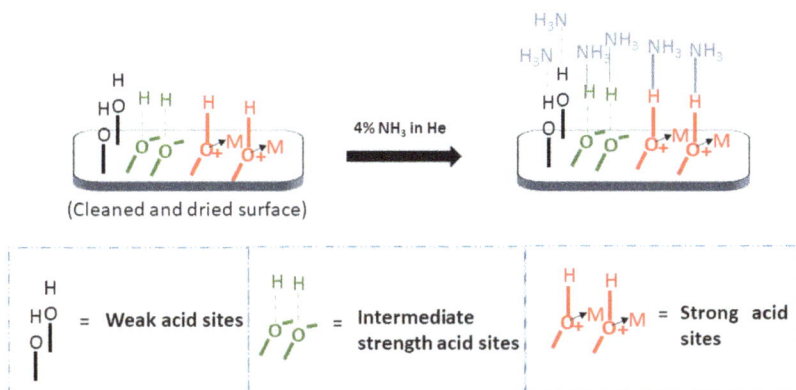

Figure 2.25: The outline of interaction of NH$_3$ with surface and surface modification.

2.3.6.2 Chemistry during de-adsorption of gas blend at increasing temperature

The outline of typical NH$_3$-desorption procedure, surface modification on increasing temperature over the catalyst surface and sense of change of conductivity in detector coil are shown in Figure 2.26. Temperature of the sample chamber is raised from 50 to 250 °C, ammonia de-adsorbs from weak acid sites and thereafter it adds to helium stream (Figure 2.26). The conductivity of NH$_3$ is higher than helium; so, conductivity of gas blend increases. As more conductive gas blend passes through the coil (fit at outlet), coil loses its temperature frequently. So, extra current is supplied from the external circuit to maintain the temperature constant. The current signal versus temperature is recorded by a detector. The outline of typical NH$_3$-desorption procedure, surface modification on increasing temperature over the catalyst surface and sense of change of conductivity in detector coil are shown in Figure 2.26.

Further, temperature of the sample chamber is raised from 250 to 550 °C. As temperature increases, ammonia de-adsorbs from intermediate-strength acid sites and adds to the helium stream. In the same way, as temperature of the sample chamber increases from 550 to 850 °C, as temperature increases, ammonia de-adsorbs from strong acid sites and adds to the helium stream. Conductivity of gas blend was increased further as more conductive NH$_3$ amount is added in the gas blend. When the more conductive gas blend is passed through the coil (fit at outlet), coil loses its temperature frequently. So, extra current is needed from the external circuit to maintain the temperature constant. The current signal versus temperature is recorded by the detector. An instrument is already calibrated by different NH$_3$ amounts (concentration) versus conductivity. By comparing, the signal intensity data can be converted into desorbed NH$_3$ amount (μmol/g catalyst per degree temperature). De-adsorbed NH$_3$ amount versus temperature graph can be plotted.

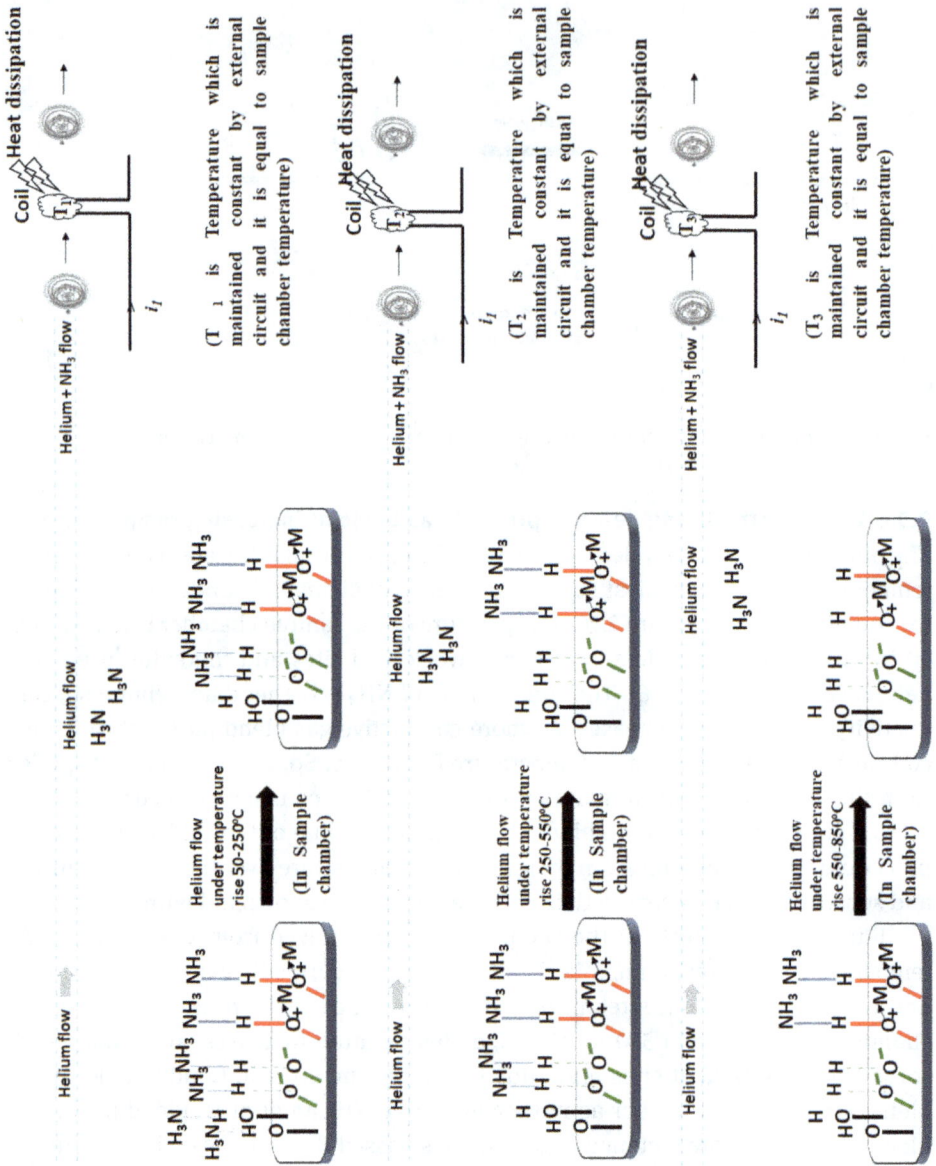

Figure 2.26: The outline of typical NH_3 desorption procedure, surface modification on increasing temperature over the catalyst surface and sense of change of conductivity in detector coil.

2.3.6.3 Analysis of NH$_3$-TPD profile

Example 16: Rawesh et al. had synthesized indium containing hexagonally ordered mesoporous silicates (In-SBA-15) by sol–gel method using the following gel composition: P123:H$_2$O:HCl:n-butanol: tetraethyl orthosilicate:In(NO$_3$)$_2$ = 6H$_2$O:0.017:200:5.4:1.325:1:0.02–0.08 [14]. NH$_3$-TPD profile of SBA-15 and indium containing SBA-15 is shown in Figure 2.27. It showed desorption peaks of NH$_3$ at low temperature (<250 °C), at intermediate temperature (<500 °C) and at high temperature (<800 °C) which were ascribed to weak acid sites, moderate-strength acid sites and strong acid sites, respectively. The catalyst having 2 mol% indium has broad density of strong acid sites.

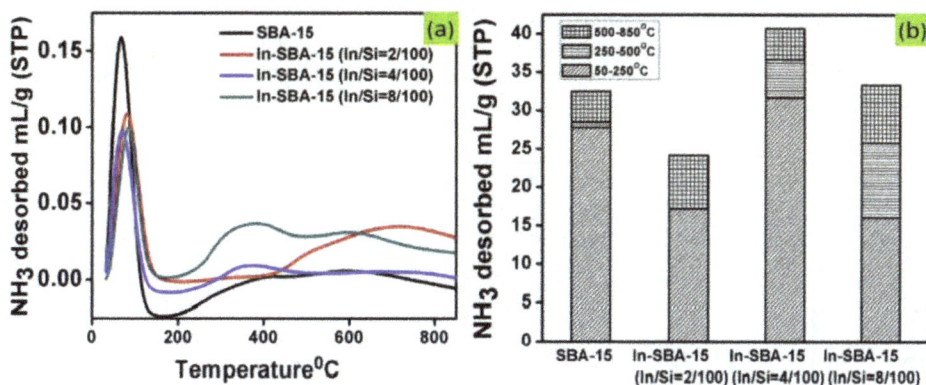

Figure 2.27: (A) NH$_3$-TPD profile of SBA-15 and indium containing SBA-15 and (B) quantitative representation of NH$_3$ at different temperature ranges for different catalyst systems. With permission from Elsevier.

Example 17: Rawesh et al. have prepared Nb$_2$O$_5$–ZrO$_2$ mixed oxide by evaporation-induced self-assembly process [15] using the following composition: NbCl$_5$:ZrOCl$_2$ 8H$_2$O:pluronic P123:EtOH = (0.5–4.5): (4.5–0.5):0.08:170 [15]. NH$_3$-TPD profile of niobia–zirconia catalyst sample is shown in Figure 2.28. The mixed oxide has broad distribution of weak, medium and strong acid sites. With the increase in zirconia up to 50%, the densities of all types of acid sites decreased. With further increase in zirconium till 70%, densities of strong acid sites kept increasing, shifting to lower temperature. Shifting of strong acid sites to lower temperature indicates accessibility of stronger acid site on lower temperature.

Example 18: Rahman et al. had prepared Gr XIII member-doped Co-incorporated hexagonally ordered mesoporous silica (M-Co-HMX; M = Al, Ga, Tl) by the sol–gel method using the following gel composition: P123:H$_2$O:n-butanol:tetraethyl orthosilicate:Co(NO$_3$)$_2$·6H$_2$O:M(NO$_3$)$_2$ = 0.017:200:5.4: 1.325:1:0.05–0.15:0.0025 [3]. NH$_3$-TPD profile of Gr XIII member-doped Co-incorporated SBA-15 sample is shown in Figure 2.29. NH$_3$-TPD profile of sample has the following information: incorporation of Al and Ga contributes to an increase in acidity in the weak range and the strong range, respectively, while Tl incorporation contributes to a decrease in acidity. An increase in Ga loading decreased the weak acidity and increased the strong acidity. Co/Ga ratio = 4 showed the maximum number of acid sites in the strong acidity range.

Example 19: Al-Mubaddel et al. had synthesized lanthana–alumina-supported Ni system by mechanical mixing of metal nitrate precursors with meso-γ-Al$_2$O$_3$, followed by drying and calcination [16]. NH$_3$-TPD profile of lanthana–zirconia-supported Ni samples is shown in Figure 2.30. The catalyst sample had a wide range of acid sites over the catalyst surface due to the acidic nature of alumina. Interestingly, with increasing La content on alumina support, acidity of alumina was thought to be neutralized and so the density of acid sites is found to decrease on addition of lanthana in alumina support. About 15 wt% lanthanum–85 wt% alumina-supported nickel catalyst had minimum acidity.

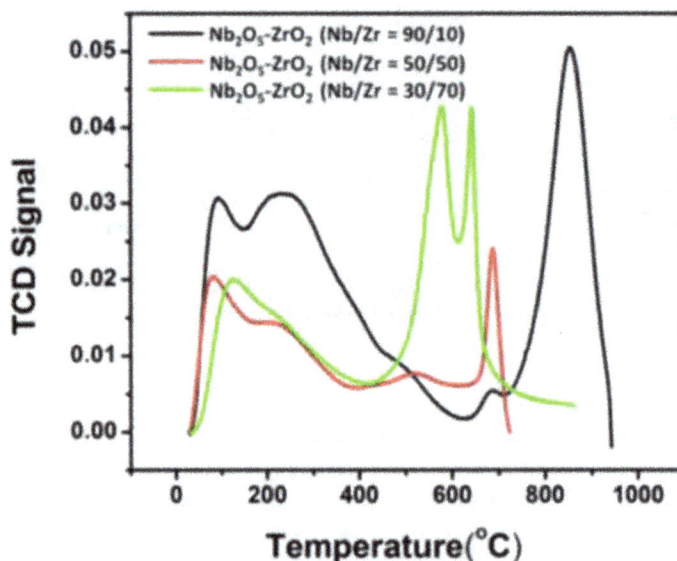

Figure 2.28: NH$_3$-TPD profile of niobia–zirconia catalyst sample. With permission from Elsevier.

2.3.7 CO$_2$-temperature-programmed de-adsorption (CO$_2$-TPD)

To illustrate the basic property of the surface, acidic probes such as CO$_2$, SO$_2$ and C$_4$H$_4$NH (pyrrole) are used under TPD techniques. CO$_2$ is acidic in nature which can be interacted with a basic surface. It adsorbed weakly with weak basic sites, moderately with intermediate-strength basic sites and strongly with strong basic sites (Figure 2.31). If surface basicity is expected to wear by oxygen, then the oxygen of surface hydroxyl (or bidentate carbonates which is formed by CO$_2$ adsorption)

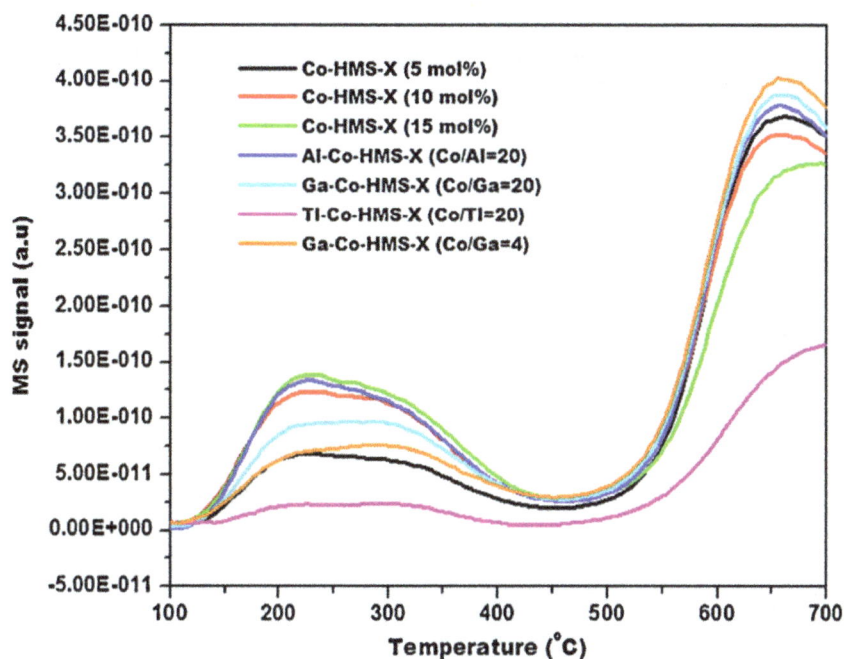

Figure 2.29: NH$_3$-TPD profile of M-doped Co-incorporated SBA-15 sample (M = Al, Ga, Tl). With permission from Elsevier.

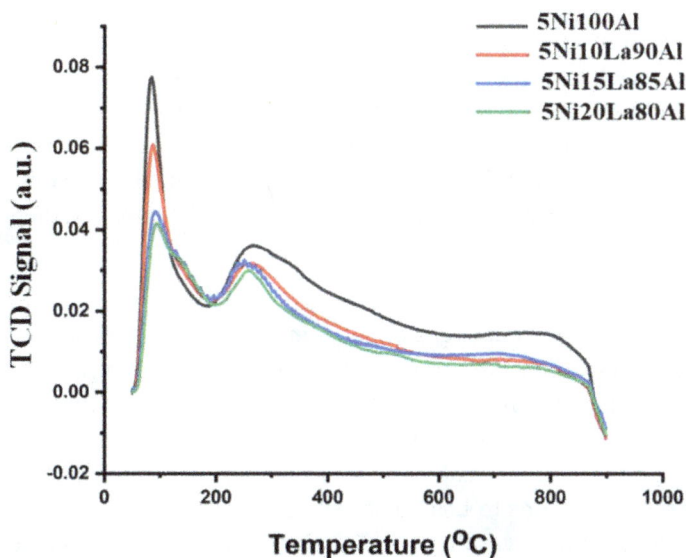

Figure 2.30: NH$_3$-TPD profile of lanthana–zirconia-supported Ni samples. With permission from Elsevier.

contributes weak basicity, surface oxygen anion contributes intermediate-strength basicity and lattice oxygen anion/oxide vacancy contributes strong basic sites. With increasing temperature, CO_2 would be de-adsorbed from the surface as per the interaction extent with the surface. During 50–200 °C, CO_2 from weak basic sites is de-adsorbed. During 200–400 °C, CO_2 from the intermediate-strength basic sites is de-adsorbed, whereas during 400–600 °C, CO_2 from strong basic sites is de-adsorbed. The high temperature (>700 °C), CO_2 desorption peaks are due to thermally stable monodentate carbonate. It is termed as super basic sites. Temperature-programmed desorption of CO_2 examines the basicity strength of surface at different temperatures.

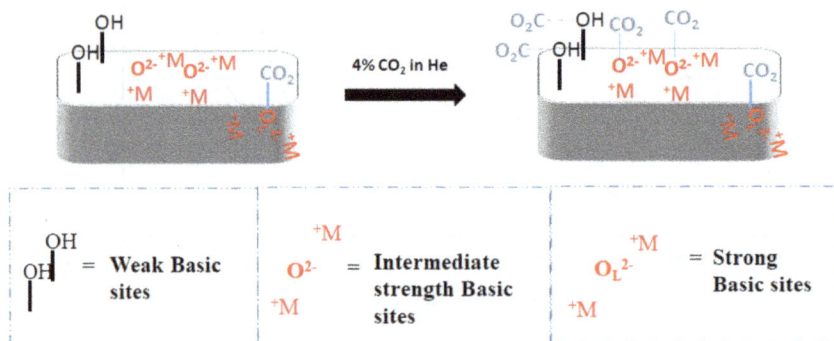

Figure 2.31: The outline of interaction of CO_2 with surface and surface modification.

2.3.7.1 The typical procedure

The sample was first degassed in flow of helium for 2 h at 200 °C. Now, 10% CO_2 in He is allowed to pass through the sample at room temperature for 30 min. Afterward, the excess CO_2 was flushed out in a flow of He for 45 min. Then the temperature-programmed desorption of ammonia (carrier gas He) was obtained by heating from ambient temperature to 1,000 °C at a temperature ramp of 10 °C/min. The thermal conductivity of gas blend is detected by TCD.

2.3.7.2 Chemistry during de-adsorption of gas blend at increasing temperature

The outline of typical CO_2-desorption procedure, surface modification on increasing temperature over the catalyst surface and sense of change of conductivity in detector coil are shown in Figure 2.32. Temperature of the sample chamber is raised from 50 to 200 °C, CO_2 de-adsorbs from weak basic sites (surface hydroxyl or bidentate carbonates) and thereafter it adds to helium stream. The conductivity of CO_2 is less than helium; so, conductivity of gas blend (CO_2 in He) decreases. As less conductive gas blend passes through the coil (fit at outlet), coil loses its temperature slowly. So, extra less magnitude of current is supplied from the external circuit to maintain the temperature constant. The current signal versus temperature is recorded by the detector. It should be noted that the conductivity of CO_2 is less than helium but higher than argon.

Further, the temperature of the sample chamber increased from 200 to 400 °C. As temperature increases, CO_2 de-adsorbs from intermediate-strength basic sites and adds to the helium stream. In the same way, as temperature of the sample chamber increases from 400 to 600 °C, as temperature increases, CO_2 de-adsorbs from strong basic sites (oxide vacancy) and adds to the helium stream. As less conductive CO_2 amount is added in the gas blend, the conductivity of gas blend was decreased further. When the less conductive gas blend is passed through the coil (fit at outlet), coil loses its temperature slowly. So, extra low magnitude of current is needed from the external circuit to maintain the temperature constant. The current signal versus temperature is recorded by the detector. An instrument is already calibrated by different CO_2 amounts (concentration) versus conductivity. By comparing, signal intensity data can be converted into desorbed CO_2 amount (μmol/g catalyst per degree temperature).

2.3.7.3 Analysis of CO_2-TPD profile

Example 20: Al-Fatesh et al. [1] have prepared 3 wt% MO_x-doped γ-Al_2O_3 solid solution (M = Si, Ti, Mo, B, W, Zr) by mechanically mixing of metal precursor salt with γ-Al_2O_3 followed by calcination [1]. γ-Alumina doped with titania and γ-alumina doped with boron trioxide were taken from Inorganic Chemistry Laboratory, Oxford University. Further, 5 wt% Ni (from nitrate precursor) was dispersed over MO_x-doped γ-Al_2O_3 by the same method discussed above. CO_2-TPD profile of 3 wt% MO_x-doped γ-Al_2O_3-supported Ni catalysts (M = Si, Ti, Mo, Zr, W, B) is shown in Figure 2.33. The *5 wt% Ni dispersed over "γ-Al_2O_3 doped with 3 wt% TiO_2"* adsorbed a large amount of CO_2 in the low and middle temperature ranges, indicating dense weak and medium basic sites on the surface. The *5 wt% Ni dispersed over "γ-Al_2O_3 doped with 3 wt% MO_x (Si, B, W)"* exhibited a wide range of CO_2 adsorption in which Si-modified sample had extended adsorption in high-temperature regions, while the 5Ni3MAl (M = Mo, Zr) showed little CO_2 adsorption.

Example 21: Al-Mubaddel et al. had synthesized lanthana–alumina-supported Ni system by mechanical mixing of metal nitrate precursors with meso-γ-Al_2O_3, followed by drying and calcination [16]. CO_2-TPD profile of lanthana–alumina-supported Ni samples is shown in Figure 2.34. The (x wt%) lanthana–(100−x) wt% alumina-supported Ni catalyst (x = 0, 5, 10, 15, 20) has CO_2 de-sorption peaks in low-temperature region (<200 °C) and intermediate-temperature region (200–400 °C). In low-temperature region, there was an intense peak about 100 °C which is attributed to CO_2 desorption from weak basic sites or surface hydroxyl and a diffuse peak below 200 °C which is attributed to CO_2 desorption as bidentate carbonate. The intermediate temperature region CO_2 desorption peak was attributed to CO_2 desorption from intermediate-strength basic sites as isolated O^{2-} surface species (surface anion). Ni stabilized over "15 wt% lanthana–85 wt% alumina" support had rich CO_2 desorption profile. It had most intense peaks at low-temperature peak (<200 °C), intermediate-temperature peak (200–400 °C) and an additional broad peak after 700 °C. The peak about 700 °C was attributed to super basic sites associated with unidentate carbonates.

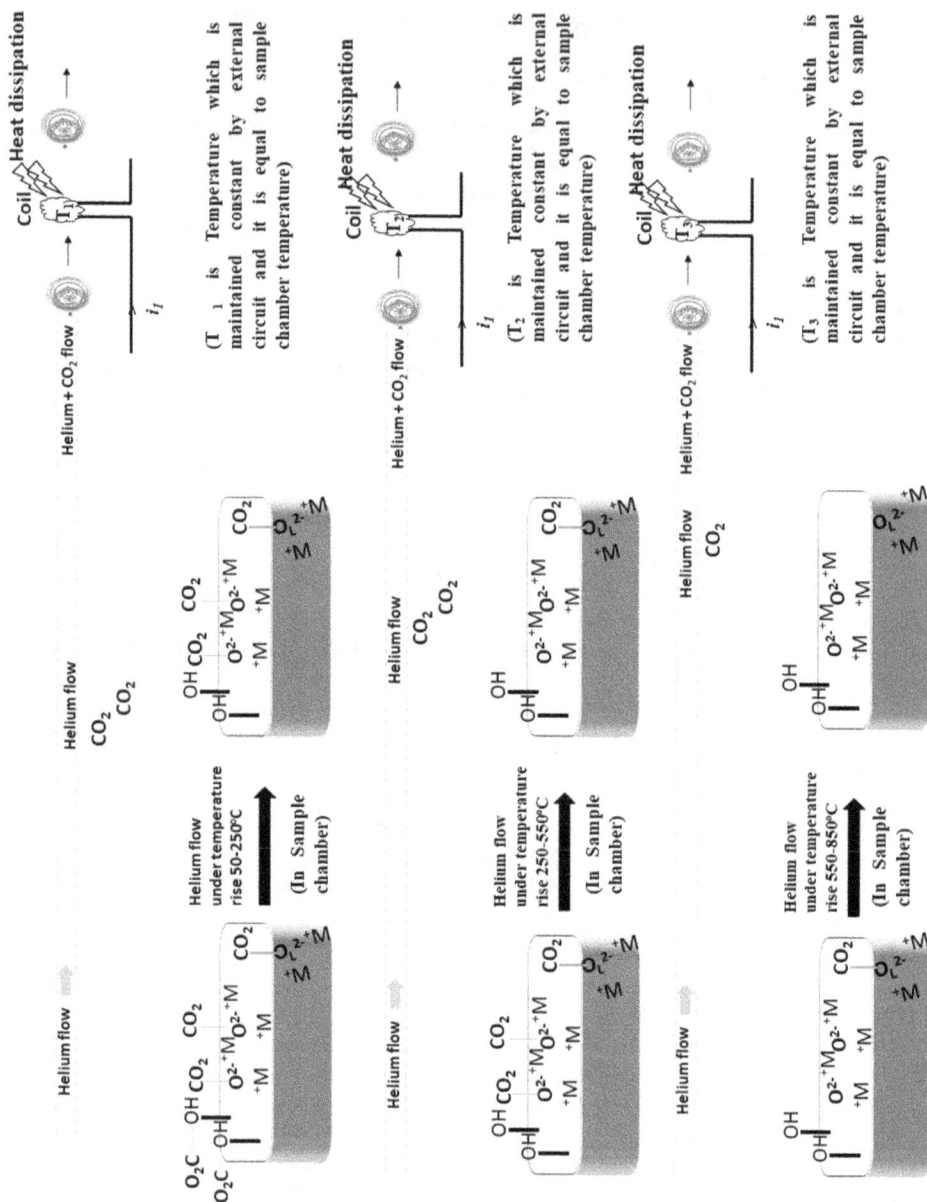

Figure 2.32: The outline of typical CO_2-desorption procedure as well as surface modification on increasing temperature over the catalyst surface and sense of change in conductivity of detector coil.

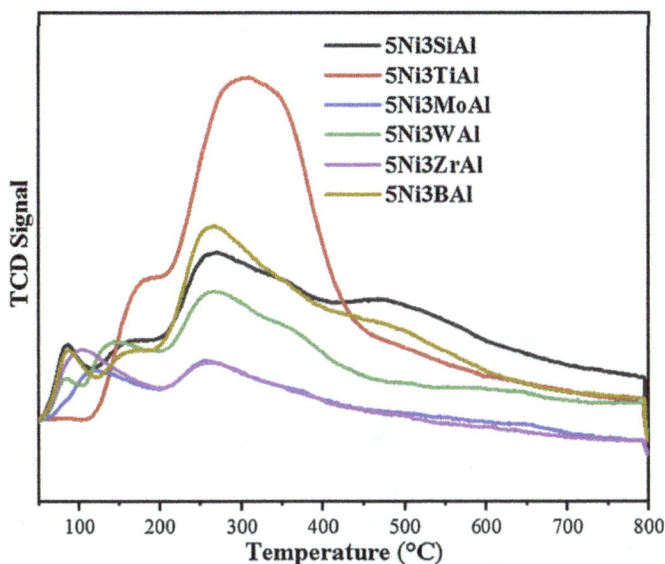

Figure 2.33: CO_2-TPD profile of 3 wt% MO_x-doped γ-Al_2O_3-supported Ni catalysts (M = Si, Ti, Mo, Zr, W, B). With permission from Elsevier.

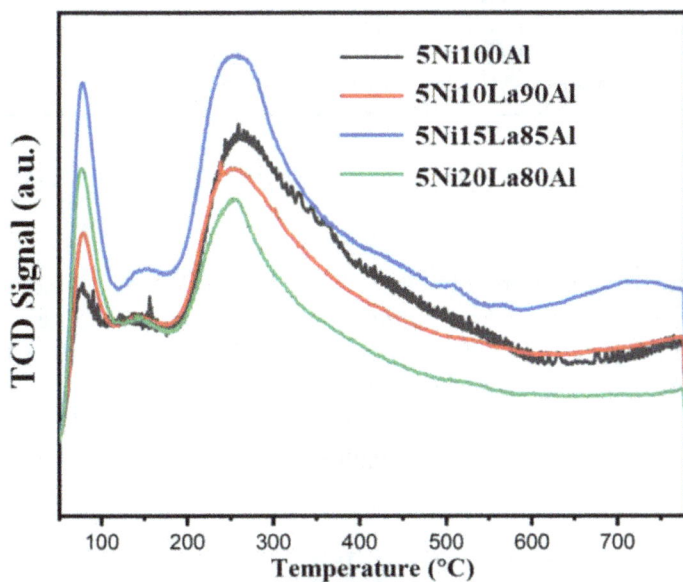

Figure 2.34: CO_2-TPD profile of lanthana–alumina-supported Ni samples. With permission from Elsevier.

Example 22: Kasim et al. prepared ceria-promoted lanthana–zirconia-supported nickel catalyst by wet impregnation of nickel nitrate precursor and ceria nitrate precursor over lanthana–zirconia [17]. CO_2-TPD profile of lanthana–zirconia support, lanthana–zirconia-supported Ni catalyst and ceria-promoted lanthana–zirconia-supported Ni catalyst is similar except 2.5 wt% ceria loading (Figure 2.35). It had desorption peaks at 80 °C, and a hump at 140 °C and 266 °C for weak basic sites (surface hydroxyl) and intermediate basic sites (surface anion), respectively. On 2.5 wt% ceria addition, peak in the range of intermediate-temperature region intensified, and an intense peak in the high-temperature region (about 650 °C) appeared. The high-temperature reduction peak about 650 °C was attributed to strong basic sites (oxide vacancy). It indicates that at this particular ceria loading, both surfaces enriched by surface anion and oxide vacancy formations were geared up.

Example 23: Rutu et al. had synthesized tungsten–zirconia-supported nickel catalyst (5Ni/WZr) and 2.5 wt% ceria-promoted tungsten–zirconia-supported nickel catalyst (5Ni2.5Ce/WZr) by the wet impregnation method using metal nitrate precursors over tungsten–zirconia support [12]. CO_2-TPD profile of fresh and spent 5Ni/WZr and 5Ni2.5Ce/WZr is shown in Figure 2.36. Tungsten–zirconia-supported Ni catalyst and 2.5 wt% ceria-promoted tungsten–zirconia-supported Ni catalyst showed CO_2 de-adsorption from weak basic sites (surface hydroxyl) in low-temperature region (<200 °C) and from intermediate-strength basic sites (surface anion) in intermediate-temperature region (200–400 °C). The CO_2-TPD of the spent catalyst system is markable. Tungsten–zirconia-supported Ni catalyst showed an intense peak about 700 °C, which was claimed to be the presence of highly thermally stable carbonates (Figure 2.36(B)). Interestingly, on 2.5 wt% ceria addition, the high-temperature peak (700 °C) shifted to 550 °C (Figure 2.36(A)). The reduction peak about 550 °C was attributed to CO_2 adsorption at strong acid sites (oxide vacancy). It indicates that on ceria addition, the stable carbonates that can block the catalyst-active sites are disappeared.

2.3.8 Cyclic experiment

The cyclic experiment is discussed further by the research work of Rutu et al. Rutu et al. had synthesized tungsten–zirconia-supported nickel catalyst (5Ni/WZr) and 2.5 wt% ceria-promoted tungsten–zirconia-supported nickel catalyst (5Ni2.5Ce/WZr) by the wet impregnation method using a metal nitrate precursor over tungsten–zirconia support [12]. Under typical examples, CO_2-TPD followed by O_2-TPO profile and TPH followed by O_2-TPO of 5Ni/WZr are shown in Figure 2.37.

2.3.8.1 CO_2-TPD followed by O_2-TPO (CO_2-TPD→ O_2-TPO)

O_2-TPO of spent tungsten–zirconia-supported Ni catalyst (5Ni/WZr) showed the peak in the range of 530–575 °C which was attributed to moderately oxidizable carbon species (β-carbon species) [12]. If O_2-TPO was carried out after CO_2-TPD, a more intense peak centered about 600 °C was found, which indicated the possible interaction of

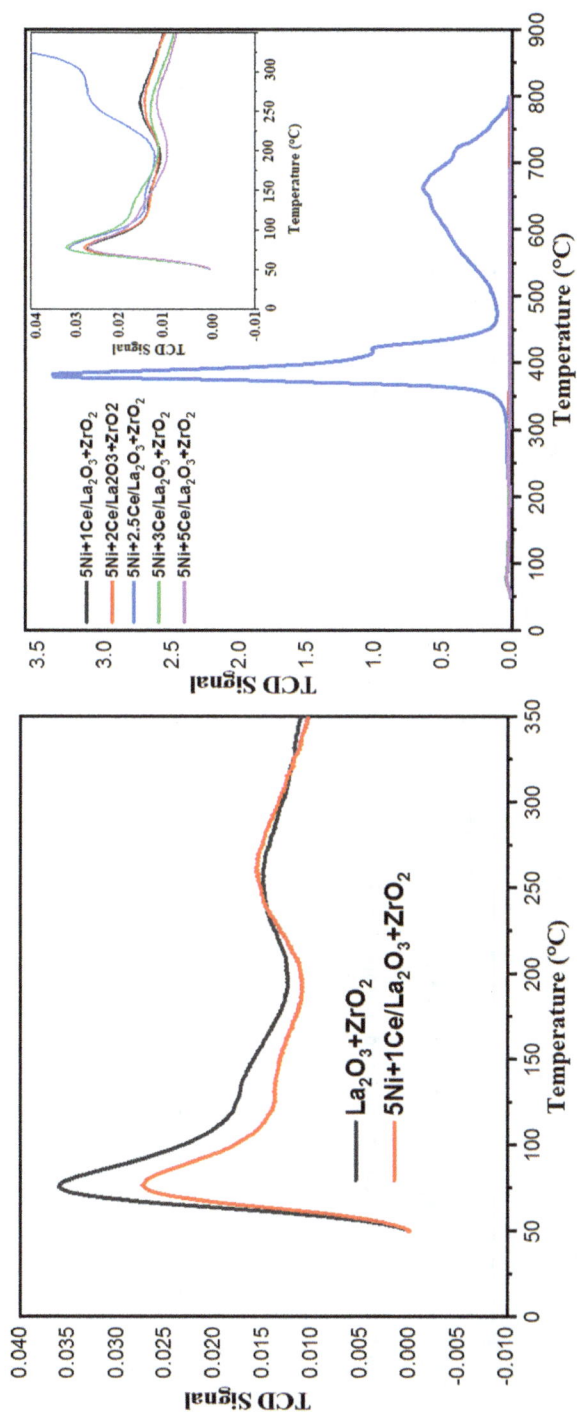

Figure 2.35: CO_2-TPD profile of lanthana–zirconia support, lanthana–zirconia-supported Ni catalyst and ceria-promoted lanthana-supported Ni catalyst. With permission from Elsevier.

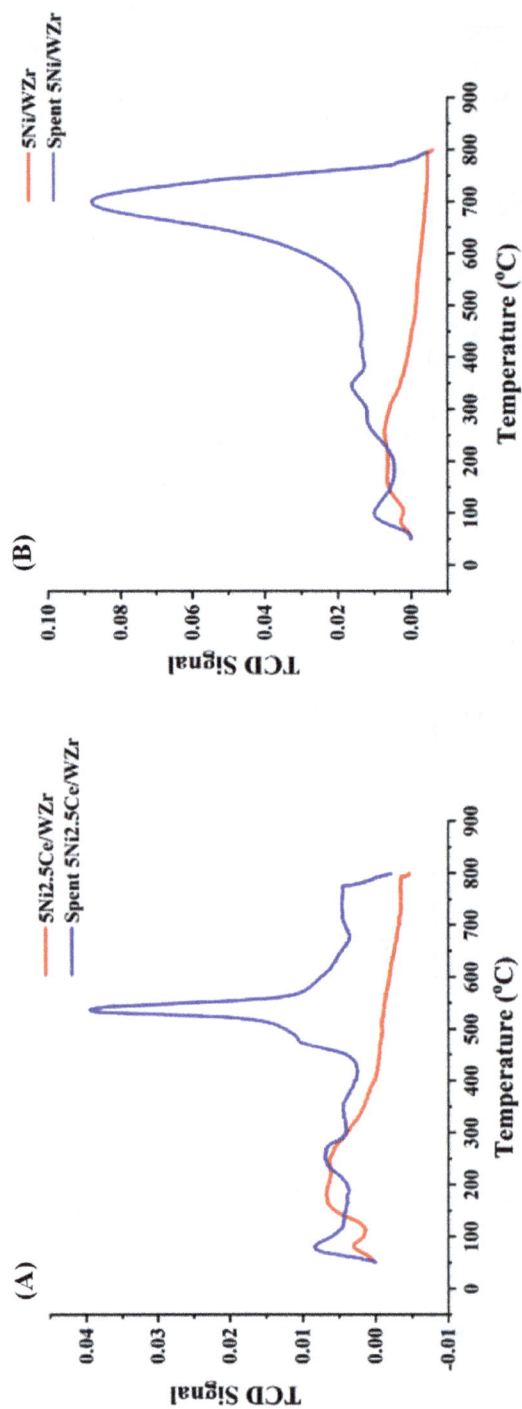

Figure 2.36: CO_2-TPD profile of fresh and spent (A) 5Ni/WZr and (B) 5Ni2.5Ce/WZr. With permission from Elsevier.

CO_2 with carbon deposit, generating oxygenated species which is further oxidized with greater intensity at higher temperature during O_2-TPO.

2.3.8.2 TPH followed by O_2-TPO (TPH → O_2-TPO)

O_2-TPO of spent tungsten–zirconia-supported Ni catalyst (5Ni/WZr) showed the peak in the range of 530–575 °C which was attributed to moderately oxidizable carbon species (β-carbon species) [12]. If O_2-TPO was carried out after TPH, the most intense peak centered about 700 °C was found, which indicated the possible interaction of H_2 with carbon deposit, generating their reduced form (higher molecular mass), which is further oxidized with greater intensity at higher temperature during O_2-TPO.

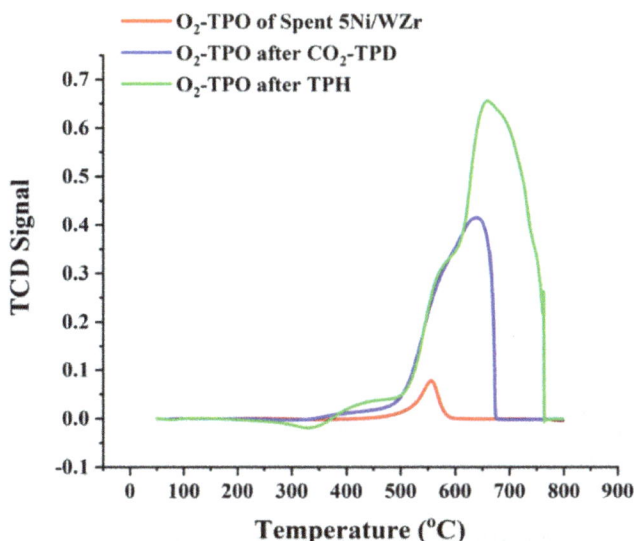

Figure 2.37: O_2-TPO profile of 5Ni/WZr, CO_2-TPD followed by O_2-TPO profile of 5Ni/WZr and TPH followed by O_2-TPO of 5Ni/WZr. With permission from Elsevier.

2.3.8.3 H_2-TPR followed by CO_2-TPD (H_2-TPR→ CO_2-TPD)

The outline of surface modification after H_2-TPR followed by CO_2 TPD is shown in Figure 2.38. CO_2-TPD profile of tungsten–zirconia-supported nickel catalyst showed CO_2 desorption peaks from lower temperature region (surface hydroxyl/weak basic sites) as well as intermediate-temperature region (surface anion/moderate-strength basic sites). But when CO_2 is carried out after H_2-TPR, CO_2 desorption peaks from intermediate-temperature region disappeared, and a high-temperature peak (~700 °C) for thermally stable carbonate appeared. It indicated that during H_2 treatment (in H_2-TPR), surface anion was removed (H_2O was formed).

CO$_2$-TPD profile of ceria-promoted tungsten–zirconia-supported nickel catalyst is not much different than the unpromoted sample. But when CO$_2$ is carried out after H$_2$-TPR, CO$_2$ desorption peaks from low-temperature region (surface hydroxyl/ weak basic sites) intensified, peak from intermediate-temperature region disappeared, and a high-temperature peak (~700 °C) for thermally stable carbonate appeared. It indicated the growth of surface hydroxyl after exposure of H$_2$ in ceria sample as well as the removal of surface anion under H$_2$ stream (H$_2$O is formed). CO$_2$-TPD profile, CO$_2$-TPD profile after H$_2$-TPR of 5Ni/WZr and 5Ni2.5Ce/WZr sample are shown in Figure 2.39.

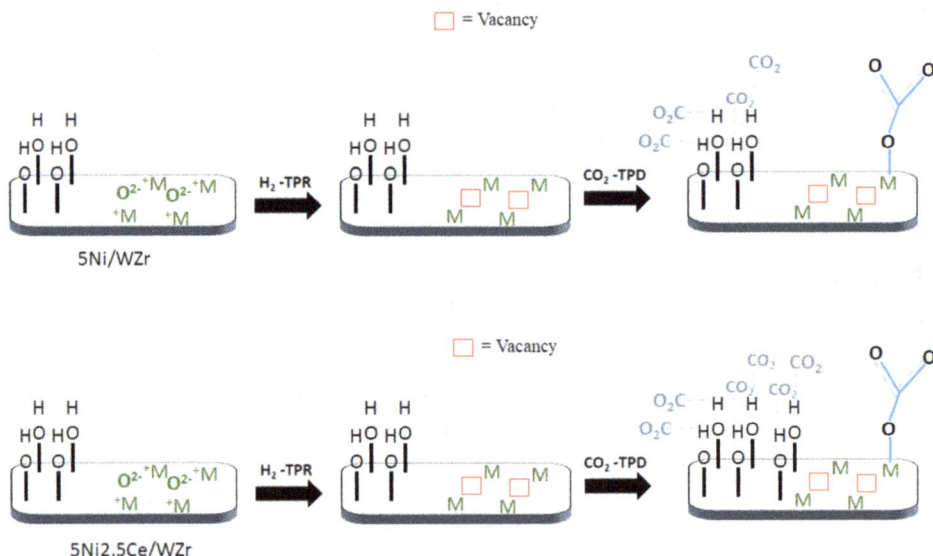

Figure 2.38: The outline of interaction of 5Ni/WZr and 5Ni2.5Ce/WZr sample under H$_2$ stream followed by CO$_2$ stream.

2.3.8.4 H$_2$-TPR followed by CO$_2$-TPD and H$_2$-TPR (H$_2$-TPR→ CO$_2$-TPD→ H$_2$-TPR)

The outline of surface modification after H$_2$-TPR followed by CO$_2$ TPD and consequent H$_2$-TPR stream are shown in Figure 2.40. The H$_2$-TPR of tungsten–zirconia-supported Ni catalyst showed diffuse reduction peaks at 245–305 °C for "nickel having weak support interaction," intense reduction peak at 485 °C for "NiO moderately interacted with support" and a broad reduction peak at 800 °C for reduction of tungsten species. About 2.5 wt% ceria-promoted tungsten–zirconia-supported Ni catalyst showed 5Ni2.5Ce/WZr catalyst had comparable reduction peaks for "nickel having weak support interaction" in low-temperature region and "NiO moderately interacted with support" in intermediate-temperature regions and a broad peak about 800 °C for reduction of tungsten species. After the reduction of surface, metal species along

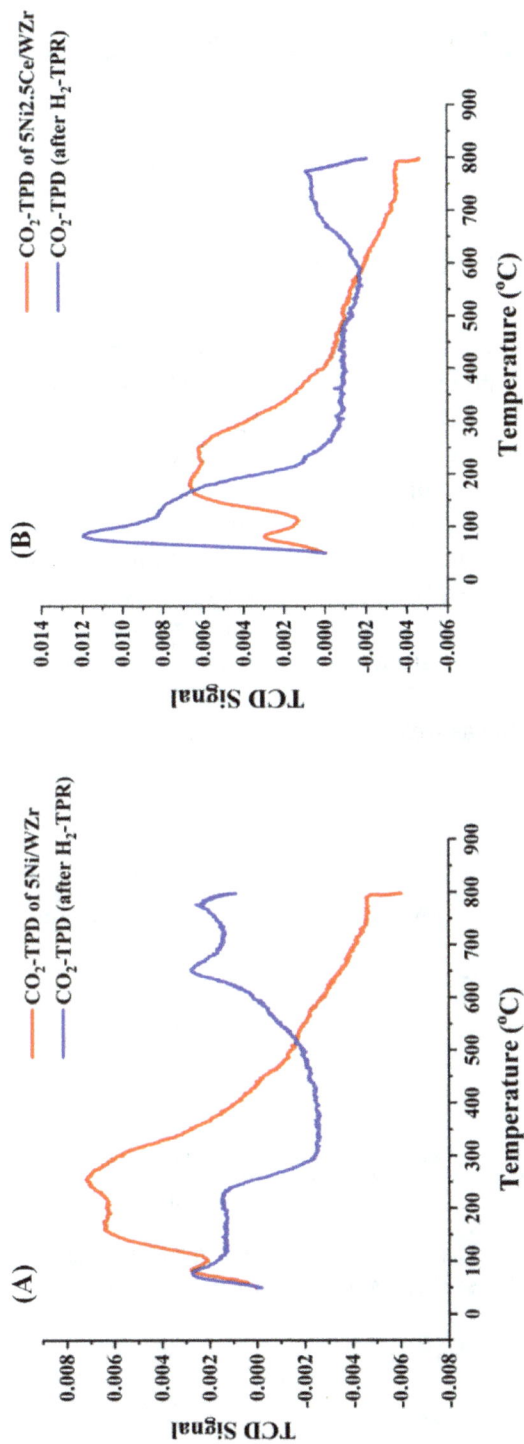

Figure 2.39: CO_2-TPD profile, CO_2-TPD profile after H_2-TPR of 5Ni/WZr and 5Ni2.5Ce/WZr sample. With permission from Elsevier.

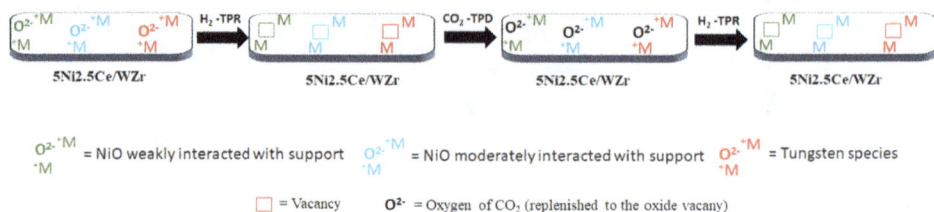

Figure 2.40: The outline of surface modification after H_2-TPR followed by CO_2 TPD and consequent H_2-TPR stream.

with oxide vacancy are formed. Further, CO_2 was carried out in continuation of H_2-TPR. If CO_2 replenished its oxygen into the oxide vacancy of surface with full extent, its surface can be reoxidized again. The reoxidation extent of surface can be checked out by carrying out H_2-TPR at the end of the cycle (H_2TPR $\rightarrow$ CO_2 TPD $\rightarrow$ H_2TPR). In the case of tungsten–zirconia-supported Ni catalyst, it was found that H_2-TPR peak is not generated up to the previous level as for fresh catalyst. It indicated that CO_2 reoxidized the surface but not up to the level of fresh catalyst. However, in the case of ceria-promoted catalyst, H_2-TPR profile was not only generated up to the native level but also it shifted toward relatively low temperature. It indicates that "CO_2 and ceria" can reoxidize the surface up to the native level and also it made the surface more reducible further. H_2-TPR profile, CO_2-TPD profile after H_2-TPR and H_2-TPR after CO_2-TPD profile as well as H_2-TPR (in sequence) are shown in Figure 2.41.

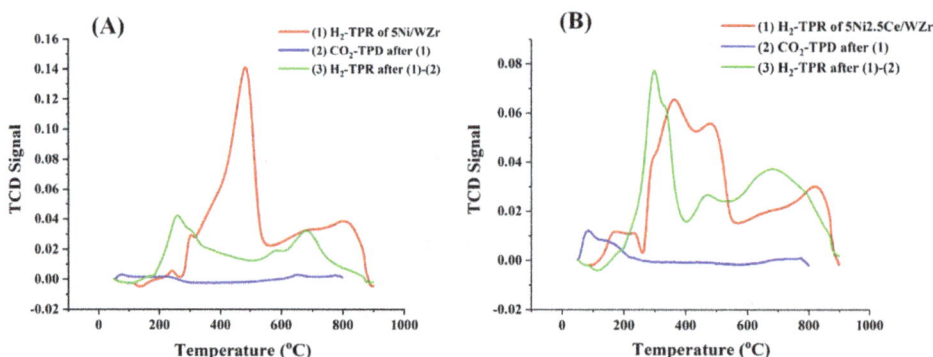

Figure 2.41: H_2-TPR profile, CO_2-TPD profile after H_2-TPR and H_2-TPR after CO_2-TPD profile and H_2-TPR (in sequence) of (A) 5Ni/WZr and (B) 5Ni2.5Ce/WZr. With permission from Elsevier.

References

[1]   Al-Fatesh, A. S., Kumar, R., Kasim, S. O., Ibrahim, A. A., Fakeeha, A. H., Abasaeed, A. E., Alrasheed, R., Bagabas, A., Chaudhary, M. L., Frusteri, F., Chowdhury, B. *Catal. Today* 2020, *348*(September 2019), 236–242.

[2]   Chu, H., Yang, L., Zhang, Q., Wang, Y. *J. Catal.* 2006, *241*(1), 225–228.

[3]   Rahman, S., Santra, C., Kumar, R., Bahadur, J., Sultana, A., Schweins, R., Sen, D., Maity, S., Mazumdar, S., Chowdhury, B. *Appl. Catal. A Gen.* 2014, *482*, 61–68.

[4]   Enjamuri, N., Hassan, S., Auroux, A., Pandey, J. K., Chowdhury, B. *Appl. Catal. A Gen.* 2016, *523*, 21–30.

[5]   Hassan, S., Kumar, R., Tiwari, A., Song, W., van Haandel, L., Pandey, J. K., Hensen, E., Chowdhury, B. *Mol. Catal* 2018, *451*September 2017, 238–246.

[6]   Al-Fatesh, A. S., Kumar, R., Fakeeha, A. H., Kasim, S. O., Khatri, J., Ibrahim, A. A., Arasheed, R., Alabdulsalam, M., Lanre, M. S., Osman, A. I., Abasaeed, A. E., Bagabas, A. *Sci. Rep.* 2020, *10*(1), 1–10.

[7]   (a) Khatri, J., Al Fatesh, A. S., Fakeeh, A. H., Ibrahim, A. A., Abasaeed, A. E., Kasim, S. O., Osman, A. I., Patel, R., Kumar R. Mol. Catal. 2021, 504, 111498.

[7]   (b) Khatri, J., Fakeeha, A. H., Kasim, S. O., Lanre, M. S., Abasaeed, A. E., Ibrahim, A. A., Kumar, R., Al-Fatesh, A. S. Int. J. Energy Res 2021, No. June, 1–14.

[8]   Schubert, M., Pokhrel, S., Thomé, A., Zielasek, V., Gesing, T. M., Roessner, F., Mädler, L., Bäumer, M. *Catal. Sci. Technol.* 2016, *6*(20), 7449–7460.

[9]   Rahman, S., Farooqui, S. A., Rai, A., Kumar, R., Santra, C., Prabhakaran, V. C., Bhadu, G. R., Sen, D., Mazumder, S., Maity, S., Sinha, A. K., Chowdhury, B. *RSC Adv.* 2015, *5*(58), 46850–46860.

[10]   Li, C., Brown, T. C. *Energy Fuels* 1999, *13*(4), 888–894.

[11]   Fakeeha, A. H., Bagabas, A. A., Lanre, M. S., Osman, A. I., Kasim, S. O., Ibrahim, A. A., Arasheed, R., Alkhalifa, A., Elnour, A. Y., Abasaeed, A. E., Al-Fatesh, A. S. *Processes* 2020, *8*(5), 522.

[12]   Patel, R., Al-Fatesh, A. S., Fakeeha, A. H., Arafat, Y., Kasim, S. O., Ibrahim, A. A., Al-Zahrani, S. A., Abasaeed, A. E., Srivastava, V. K., Kumar, R. *Int. J. Hydrogen Energy* 2021, 46, 25015–25028.

[13]   Patel, R., Fakeeha, A. H., Kasim, S. O., Sofiu, M. L., Ibrahim, A. A., Abasaeed, A. E., Kumar, R., Al-Fatesh, A. S. *Mol. Catal* 2021, *510*April, 111676.

[14]   Kumar, R., Shah, S., Bahadur, J., Melnichenko, Y. B., Sen, D., Mazumder, S., Vinod, C. P., Chowdhury, B. *Micropor. Mesopor. Mater* 2016, *234*, 293–302.

[15]   Kumar, R., Ponnada, S., Enjamuri, N., Pandey, J. K., Chowdhury, B. *Catal. Commun.* 2016, *77*, 42–46.

[16]   Al-mubaddel, F. S., Kumar, R., Lanre, M., Frusteri, F., Aidid, A., Kumar, V., Olajide, S., Hamza, A., Elhag, A., Osman, A. I., Al-Fatesh, A. S. *Int. J. Hydrogen Energy* 2021, *46*(27), 14225–14235.

[17]   Kasim,S. O., Al-Fatesh, A. S., Ibrahim, A. A., Kumar, R., Abasaeed, A. E., Fakeeha, A. H., Int J Hydrog Energy 2020, 45, 33343–33351.

3 Fourier-transform infrared spectroscopy (FT-IR)

3.1 Background

The charge distribution about atoms in a molecule is not symmetric. If we consider two atoms about a bond (as HCl), it can be said that the charge density about atoms is not similar. The magnitude of charge difference and distance between two atoms is known as a *dipole* (dipole = charge difference × bond length). Again, atoms of molecules are not a fixed structure; they are vibrating about the mean position (r_o) with certain frequency (ω) (known as natural frequency). During vibration again, the dipole keeps on changing. But as vibration is periodic (repeating after a time interval), the *change in dipole* is also periodic. Now, if the infrared (IR) radiation is passed through the molecule, the oscillating electrical component of IR radiation interacts with the oscillating dipole vector of a molecule. If the frequency of radiation matches with the natural vibrational frequency of the molecule, there occurs a net transfer of energy from the radiation source to the molecule (absorption of radiation by the molecule) by which the molecule will attain the next level of vibration ($\Delta v = \pm 1$, where v is the vibrational quantum level).

For an ideal condition, an oscillator (harmonic oscillator) should obey *Hooke's law,* which states that a system displaced from equilibrium responds with a restoring force whose magnitude is proportional to the displacement (Figure 3.1). $F = -K r_o$; K is the "force constant" and r_o is the distance of the mean position. The frequency is derived as $\omega = (1/2\pi)\sqrt{K/\mu}$; $\mu = m_1 m_2/(m_1 + m_2)$; μ is the reduced mass. As vibrational energy is quantized, the vibrational energy of the molecule is shown by the Schrodinger wave equation $E = \left(v + \dfrac{1}{2}\right)h\omega$, where v is the vibrational quantum number and $v = 0, 1, 2, \ldots$. Clearly for a harmonic oscillator, the molecule absorbs IR absorption and attains the next level of vibration. So, a selection rule for the allowed transition is $\Delta v = \pm 1$.

3.1.1 Harmonic oscillator

The vibration energy of a molecule of the harmonic oscillator is

$$E = \left(v + \frac{1}{2}\right)h\omega; \quad v = 0, 1, 2, \ldots$$

At the ground state, $v = 0$; so, the vibrational energy of the molecule of the harmonic oscillator is

https://doi.org/10.1515/9783110656480-003

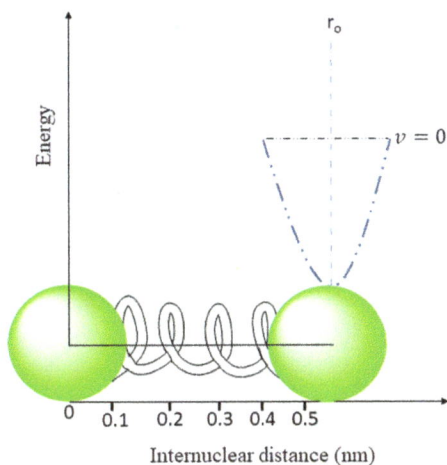

Figure 3.1: Diatomic molecule vibration under Hooke's law.

$$E = \left(\frac{1}{2}\right) h\omega$$

After interaction with the IR wave, "v" rises from 0 to 1 ($\Delta v = \pm 1$ condition is satisfied) and energy at the excited level is equal to

$$E = \left(1 + \frac{1}{2}\right) h\omega$$

The difference of vibrational levels attained by the molecule is equal to the absorption of the IR band by the molecule. This band is known as the fundamental band:

$$\Delta E = \left(1 + \frac{1}{2}\right) h\omega - \left(\frac{1}{2}\right) h\omega = h\omega$$

At the excited state $v = 1$, population is negligible on room temperature. So further excitation from $v = 1$ to $v = 2$ (for satisfying $\Delta v = \pm 1$) has very low intensity and not observable. However, at high sample temperature, population is appreciable and the absorption band is observable. This transition is known as the hot band. Overall, either fundamental band or hot band, it can be said that the energy level of the harmonic oscillator is equally spaced (Figure 3.2).

However, a system does not remain ideal, and the above harmonic oscillator model is fit only for the diatomic molecule. For anharmonic expression, the energy term is given by the Taylor series (discussed in the next section).

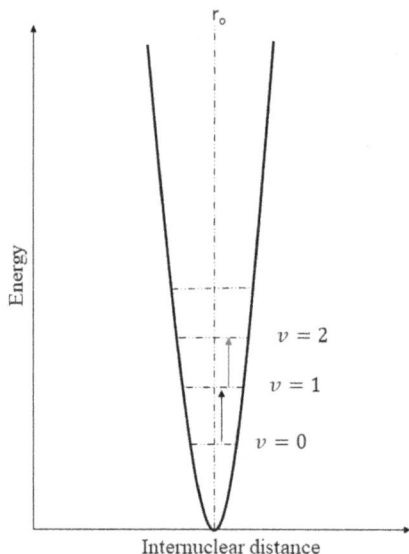

Figure 3.2: Potential energy versus internuclear distance plot for the harmonic oscillator.

3.1.2 Anharmonic oscillator

The vibration energy of the molecule of anharmonic oscillator is

$$E = \left(v + \frac{1}{2}\right)\hbar\omega - \left(v + \frac{1}{2}\right)^2 \hbar\omega_x + \left(v + \frac{1}{2}\right)^3 \hbar\omega_y + \cdots$$

where $\omega >> \omega_x >> \omega_y$; in the same way, the higher term has a negligible value and it can be neglected. The first term of energy expression is even prominent and so $\Delta v = \pm 1$ is still most predominant, whereas weaker transition $\Delta v = \pm 2, \pm 3, \ldots$ can occur. The energy level of anharmonic oscillator does not remain equally spaced (Figure 3.3).

As, for example, $\Delta v = 2$, transition occurs at twice the frequency of the fundamental transition but with low intensity. It is designated as $2v_1$ (first overtone). Same $\Delta v = 3$ transition occurs at thrice the frequency of the fundamental transition but with more low intensity. It is designated as $3v_1$ (second overtone) and so on.

So now for a polyatomic molecule at normal temperature, it may have different fundamental frequencies (at v_1, v_2, v_3, ...) and their overtone bands (at $2v_1$, $3v_1$, $2v_2$, $3v_2$, ...). The combination band (at $v_1 + v_2$; $2v_1 + v_2$) and difference band ($v_1 - v_2$; $2v_1 - v_2$) are also observed with weak intensities.

Now, the question is that how many vibration peaks are we getting to form a molecule composed of "N" atoms. The answer is embedded in the molecular vibration motions.[b] If a molecule has "N" number of atoms, it has total $3N$ degrees of freedom. Separately, it has three translational degrees of freedom, three rotational degrees of freedom and rest $3N-6$ vibrational degrees of freedom. That means for a nonlinear

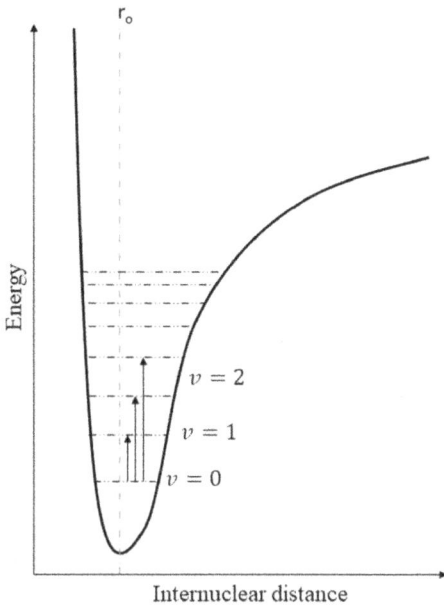

Figure 3.3: Potential energy versus internuclear distance plot for anharmonic oscillator.

molecule, $3N–6$ vibration band is expected. For a linear molecule, rotation about the molecular axis is not possible. So, a linear molecule has three translational degrees of freedom, two rotational degrees of freedom and rest $3N–5$ vibrational degrees of freedom. Again, vibration bands are classified into stretching band and bending band. A nonlinear molecule having "N" number of atoms has $N − 1$ number of stretching bands and rest $(3N–6)–(N–1)$ bending bands. For a linear molecule, banding bands are $(3N–5)–(N–1)$, as we know that more energy is required in stretching than bending. So, stretching frequency is higher than the bending frequency. Again among stretching, asymmetric stretching vibration has higher frequency than symmetric stretching vibration. If the molecular weight is comparable, the force constant[c] about a bond gives us the preliminary idea of vibrational frequency position. But when the molecular weight about a bond had a big difference, the reduced mass[d] decides the vibrational frequency position. There are many factors[e] (electronic effect and hydrogen bonding) that affect the vibrational frequency shift from the standard one.

3.2 Instrumentation and working principle

The schematic diagram of IR spectrometer is shown in Figure 3.4. Nernst glower and incandescent wire are used as IR radiation sources. Further, to separate near-IR region (wavelength; λ = 1–2.5 μm) beam, mid-IR region (λ = 2.5–25 μm) beam and far-IR region beams (λ = 25–400 μm), lithium fluoride crystal, sodium chloride crystal, potassium bromide crystal (or cesium chloride) are utilized, respectively. We are interested in the mid-IR region (λ = 25–400 μm and frequency 400–4,000 cm^{-1}).

Now, the mid-IR beam is split into two beams: one beam strikes and reflects back from the fixed mirror and another beam strikes and reflects back from the movable mirror. Both beams meet together and due to the movable mirror, the path difference of both beams are not the same. One beam marches ahead than the another beam and so if they meet, they create constructive and destructive patterns. As well as the mirror moves, a continuous range of constructive patterns gets generated, and quickly a complete IR spectrum range gets generated. In this way, the entire range of IR spectrum passes through the sample simultaneously with very high resolution ($\leq$0.001 cm^{-1}).

If the frequency of radiation matches with the natural vibrational frequency of the sample molecule, then there occurs a net transfer of energy from the radiation source to a molecule by which the molecule will attain the next level of vibration. It is called absorption of radiation by the molecule. The remained radiation (not adsorbed) comes out of a sample cage (called transmitted radiation) and reaches the detector. Many types of detectors like thermocouples, bolometers, thermistors and Golay cell are used. A detector converts radiation into an electrical signal. Now, data undergoes analogue to digital conversion by a circuit which can be analyzed over the display.

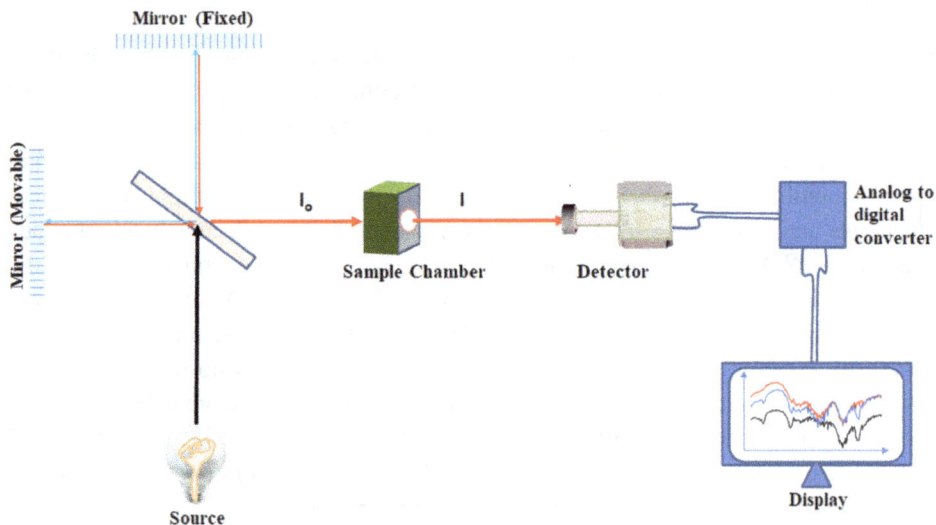

Figure 3.4: Schematic diagram of infrared spectrometer.

Overall, IR radiation of "I_o" intensity is glanced over the sample and "I" intensity is transmitted from the sample and reached the detector. The absorbance (A) is studied by monitoring both the incident intensity (I_o) and transmitted intensity of radiation (I_o) as $A = \log (I_o/I)$. Transmittance (T) is defined by $T = I/I_o$. So, the percentage transmittance can be shown as

$$T\% = \frac{I}{I_o}.100$$

Putting log on both sides, $\log (T\%) = \log(I/I_o) + \log 100$

$$\log (T\%) = -\log\left(\frac{I_o}{I}\right) + 2$$

After substituting the absorbance term, $\log (T\%) = -A + 2$

$$A = 2 - \log(T\%)$$

In this way, interconversion of intensity data either transmittance (T) or absorbance (A) can be carried out by the formula $A = 2 - \log(T\%)$. IR spectrum is presented as "transmittance versus frequency" or "absorbance versus frequency."

3.3 Glossary

[a]**Derivation of frequency:** If a mass "m" is displaced by the distance y along the axis of a spring from its equilibrium position by a force, the restoring force is developed inside the spring that is proportional to displacement (y), $F = -k\,y$, where k is known as the spring constant. If this equation equalizes with the Newton force $(F = ma$; m is mass and a is acceleration), it becomes $ma = -ky$. After writing acceleration in the form of second derivative of distance, it becomes

$$m\frac{\delta^2 y}{\delta t^2} = -ky$$

The solution of this equation will be $y = A \sin\left(\sqrt{k/m}\,t\right)$, where A is the amplitude of vibration which is the maximum value of displacement. After comparing this equation with the sinusoidal function $y = A \sin 2\pi\omega t$, it can be written as

$$A \sin\left(\sqrt{\frac{k}{m}}\,t\right) = A \sin 2\pi\omega t$$

$$\sqrt{\frac{k}{m}} = 2\pi\omega$$

$$\omega = \frac{1}{2\pi}\sqrt{\frac{k}{m}}$$

For a diatomic molecule, the mass of both ends vibrate and then in place of mass, reduced mass term is used. The reduced mass (μ) is defined as $\mu = m_1 m_2 / (m_1 + m_2)$ and now the frequency term can be written as follows:

$$\omega = \frac{1}{2\pi} \sqrt{\frac{k}{\mu}}$$

[b]**Vibration motions:** stretching and bending are two modes of vibration. Different vibration modes are picturized in Table 3.1. In stretching vibration bond length of atoms are changing (with respect to the central atom) without affecting the bond angle, whereas in bending, the vibrational bond angle changes without affecting the bond length. Again, stretching vibration can be categorized into symmetric stretching and asymmetric stretching. In symmetric stretching, atoms at both sides (about central atom) are either stretching or compressing together. In asymmetric stretching, if an atom at one side (about the central atom) is compressed, then an atom at another side (about the central atom) has to be stretched. Bending vibration is categorized into in-plane bending and out-plane bending. In in-plane bending, atoms (under motion) and central atom (about which bending takes place) lie in the same plane during the whole motion. If two atoms attached to the central atom move toward each other or move far to each other (movement in the opposite direction) and bond angle changes, then this bending is known as in-plane scissoring bending. Again, two atoms attached to the central atom move in the same direction such that the bond angle changes, and this bending is known as in-plane rocking bending. In out-of-plane bending, atoms (under motion) and the central atom (about which bending takes place) do not lie in the same plane during the whole motion. The central atom is assumed to remain constant in-plane and two atoms (attached to central atom) move either above or below the plane together. This motion is out-of-plane wagging vibration. If the central atom remains in one plane and among two atoms (attached to the central atom), one atom moves above the plane and another moves below the plane. This motion is out-of-plane twisting vibration.

[c]Force constant and vibrational frequency: If mass is comparable, then the next frequency deciding factor is force constant. The force constant (K) of a triple bond ($C \equiv C$) and a double bond ($C = C$) are thrice and double of the single bond ($C–C$). So, stretching frequencies ($\omega = (1/2\pi)\sqrt{K/\mu}$) will be in the order of $\sqrt{3}:\sqrt{2}:1$, respectively; and the expected stretching frequencies will be 2,100 cm^{-1}, 1,600 cm^{-1} and 1,200 cm^{-1}, respectively.

[d]Reduced mass and vibrational frequency: In "A–H" bond if A is low atomic weight atom like O, N and C, the reduced mass ($\mu = m_1 m_2 / (m_1 + m_2)$) is very low. So, stretching frequency ($\omega = (1/2\pi)\sqrt{K/\mu}$) is very high in the order of 3,000 cm^{-1}. In the "M–Y" bond if M is a high atomic weight atom or metal, the reduced mass ($\mu = m_1 m_2 / (m_1 + m_2)$) is very high. So, the stretching frequency ($\omega = (1/2\pi\sqrt{K/\mu})$ is very low, that is, <900 cm^{-1}.

Table 3.1: Different modes of stretching and bending vibration.

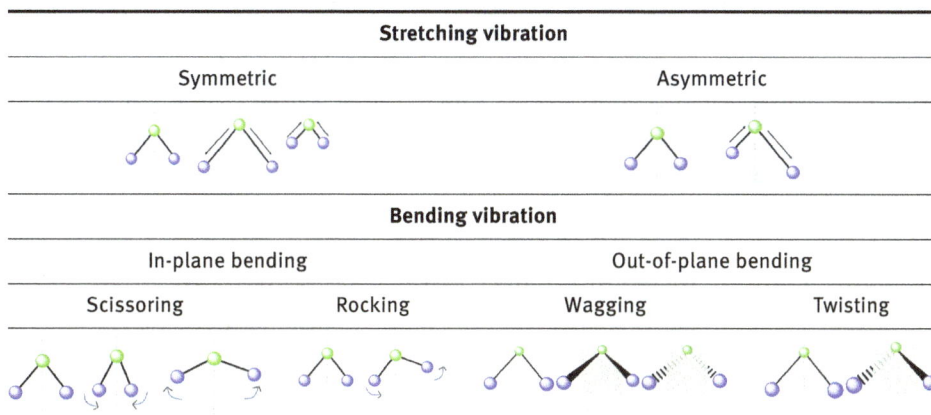

Stretching vibration	
Symmetric	Asymmetric

Bending vibration			
In-plane bending		Out-of-plane bending	
Scissoring	Rocking	Wagging	Twisting

[e]Factors affecting vibrational frequency shift: Different factors responsible for vibrational frequency shift are picturized in Figure 3.5 and tabulated in Table 3.2 with a suitable example. Due to the electronic effect (inductive, resonance effect and field effect), the electronic cloud density between atoms increased/decreased which amends the force constant/vibration frequency accordingly. Let us take an example of C = O group (in formaldehyde) which had a fundamental vibrational frequency at 1,750 cm^{-1}. Introduction of alkyl group about C = O causes positive induction (+I effect) at the carbon of carbonyl. It creates an electron-rich region at the carbon of carbonyl and weak electron density between C–O. Simply, the bond constant between C–O weakens, which causes a relatively less force constant or relatively lower frequency than C = O in formaldehyde. Again, introduction of electronegative atom about C = O (chloroacetone) makes negative induction (–I effect) at the carbon of carbonyl. It creates an electrodeficient center at the carbon of carbonyl, and the lone pair of oxygen (of carbonyl) is pulled in between C–O bond. Simply, a bond between C–O strengths causes relatively more force constant or relatively higher frequency than C = O in acetone. Due to delocalization of conjugated π bond toward the carbon of carbonyl (+R effect), the carbon of carbonyl becomes electron rich, electron density between C–O weakens and force constant/vibrational frequency relatively shifted to the lower value. Similarly, as delocalization of π electrons is far from the carbonyl (–R effect), the carbon of carbonyl becomes electrodeficient and the lone pair of oxygen (of carbonyl) is delocalized between C–O. It causes an increase in the bond strength/force constant/frequency of C–O bond than C = O in acetone. Again, in ortho-substituted compound (*ortho*-haloacetophenone), lone pair of two atoms at the substituent can influence each other through space. Halo group and oxygen of aceto-group of *ortho*-haloacetophenone influence each other through space such that the carbonyl group gets out of plane of the aromatic ring. So, +R effect from ring π electron to the carbonyl is halted. Hence, it had good electron density between C–O, which causes higher force

constant/vibrational frequency. Lastly, hydrogen bonding alters the force constant across the bond. The strength of hydrogen bonding is maximum when the proton donor group and the axis of lone pair orbital are colinear. Due to hydrogen bonding, the bond electron density between two atoms weakens, which causes lower force constant/lower vibrational frequency with respect to the compound having no hydrogen bonding, as amine in the condensed phase has higher vibrational frequency of N–H bond than the diluted one.

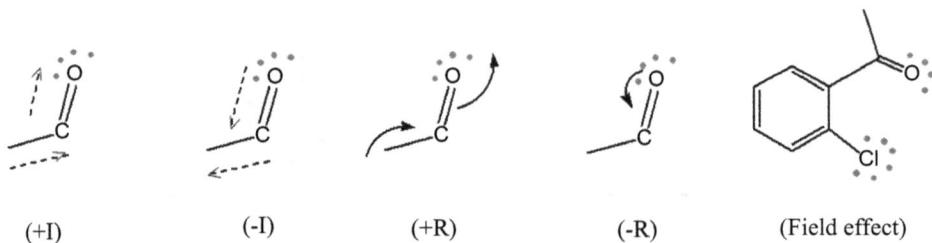

Figure 3.5: Different factors responsible for vibrational frequency shift.

Table 3.2: Vibrational frequency shift of molecules that are different from each other by means of $+I/-I$ effect, $+R/-R$ effect, field effect and H-bonding.

		$+I$ effect →			
HCHO	$1{,}750$ cm^{-1}	CH_3CHO	$1{,}745$ cm^{-1}	CH_3COCH_3	$1{,}715$ cm^{-1}
		$-I$ effect →			
CH_3COCH_3	$1{,}715$ cm^{-1}	$ClCH_2COCH_3$	$1{,}725$ cm^{-1}	$Cl_2CHCOCH_3$	$1{,}740$ cm^{-1}
		$+R$ effect →			
CH_3COCH_3	$1{,}715$ cm^{-1}	$ArCOCH_3$	$1{,}693$ cm^{-1}	$p-NH_2ArCOCH_3$	$1{,}670$ cm^{-1}
		$-R$ effect →			
CH_3COCH_3	$1{,}715$ cm^{-1}	$p-NO_2ArCOCH_3$	$1{,}770$ cm^{-1}		
		Field effect →			
$ArCOCH_3$	$1{,}693$ cm^{-1}	$o-XArCOCH_3$	$>1{,}700$ cm^{-1}		
		Hydrogen bonding →			
-NH (dilute) (amine)	$3{,}500$ cm^{-1}	-NH (condense) (amine)	$3{,}300^{-1}$		

3.4 Analysis of FTIR profile

Example 1: Al-Fatesh et al. [1] prepared 5 wt% Ni dispersed over "γ-alumina doped with 3 wt% MO_x (M = Ti, Si, Mo, W)" by mechanical mixing of nickel nitrate precursor salt with "γ-alumina doped with 3 wt% SiO_2" followed by calcination [1]. The catalyst was abbreviated as 5Ni3MAl (M = Ti, Mo, W, Si). The IR spectra of Al_2O_3, MO_x–Al_2O_3 and 5Ni3MAl (M = Ti, Mo, W, Si) catalyst samples are shown in Figure 3.6. The IR spectra of the catalyst samples showed a clear shift of Ni–O peak from 433 cm^{-1} (for titanium-modified catalyst) to 424 cm^{-1} (for molybdenum-modified catalyst) and 403 cm^{-1} (for silica-modified catalyst). That reflects the presence of "free NiO" species in titania-modified Al_2O_3-supported nickel catalyst; "NiO species interacted with the modifier" in molybdenum-modified Al_2O_3-supported nickel catalyst and "NiO species strongly interacted with the support/modifier" in silica-modified Al_2O_3-supported nickel catalyst. Tungsten-modified Al_2O_3-supported nickel catalyst samples had a widely interacted NiO species as it showed the stretching vibration of NiO_6 polyhedral.

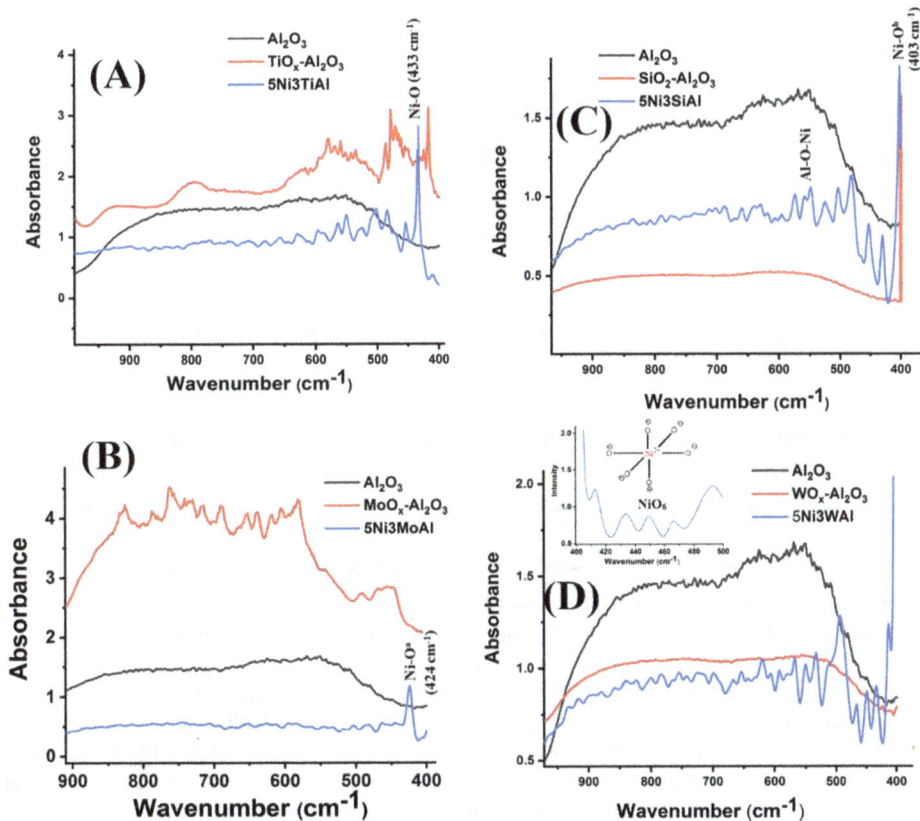

Figure 3.6: The infrared spectra of Al_2O_3, MO_x–Al_2O_3 and 5Ni3MAl (M = Ti, Mo, W, Si). With permission from MDPI.

Example 2: Rutu et al. prepared "x wt% yttria zirconia ($x = 0$–20)" support by mechanical mixing. Further, 5 wt% Ni was impregnated over yttria–zirconia support [2]. The catalyst was abbreviated as Ni–xY–ZrO$_2$ ($x = 5$–20). The IR spectra of ZrO$_2$, Ni–ZrO$_2$ and Ni–xY–ZrO$_2$ ($x = 5$–20) catalyst samples are shown in Figure 3.7. Over ZrO$_2$ sample, stretching vibration of Zr–O was observed at IR band at 748 and 500 cm^{-1}. The earlier band intensity got decreased after incorporating Y$_2$O$_3$ in ZrO$_2$ support, which indicated perturbation of Zr–O band due to the introduction of heteroatom. In all catalyst samples, the IR band for stretching vibration of OH at 3,425 cm^{-1}, for bending vibration of OH at 1,625 cm^{-1}, for vibration of physically adsorbed CO$_2$ gases at 2,334 cm^{-1}, for absorption of HCO$_3$ at 1,050 cm^{-1}, for bidentate carbonate at 1,087 cm^{-1} and for unidentate carbonate bands at 1,384 cm^{-1} and 1,517–1,530 cm^{-1}, for stretching vibration of C–H (of bicoordinated format species) at 2,850 cm^{-1} and for the combination of asymmetric stretching of COO and bending vibration of C–H bond at 2,925 cm^{-1} were noticed. After yttria loading, the band intensity for unidentate carbonate deepened. Specially at 15% Y loading, bands for unidentate carbonate, bidentate carbonate and formate species were found prominently.

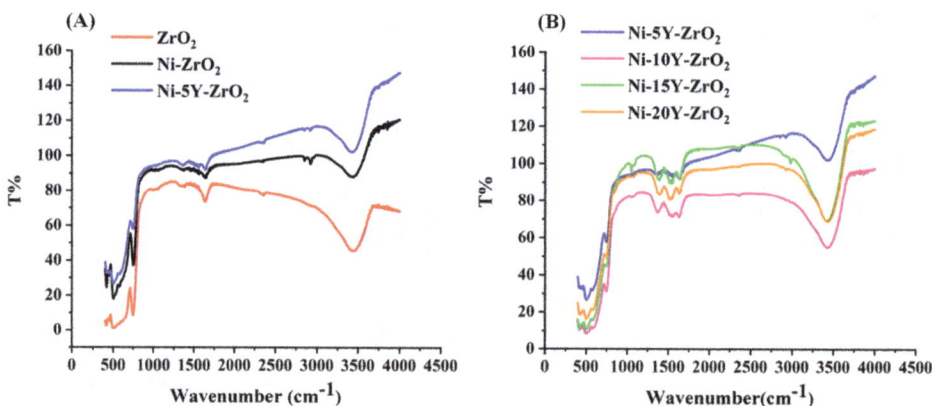

Figure 3.7: The infrared spectra of ZrO$_2$, Ni-ZrO$_2$ and Ni-xY-ZrO$_2$ ($x = 0$–20) catalyst samples. With permission from Elsevier.

Example 3: Al-Fatesh et al. prepared 5 wt% Ni dispersed over "x wt% 20 lanthana–$(100$–$x)$ wt% alumina ($x = 0, 20, 15, 20$)" by mechanical mixing of nickel nitrate, lanthana nitrate and meso-γ-alumina followed by calcination [3].

The sample was abbreviated as 5NixLa$(100$–$x)$Al ($x = 10,15, 20$). The IR spectra of 5Ni100Al, 5NixLa$(100$–$x)$Al ($x = 10,15, 20$) catalyst samples are shown in Figure 3.8. All alumina-supported catalyst had all three characteristic bands at 1,635, 1,517 and 1,382 cm^{-1} for γ-Al$_2$O$_3$. Apart from these IR bands for physically adsorbed CO$_2$ at 2,349 cm^{-1}, for bidentate CO$_2$ chemisorbed species at 1,730 and 1,270 cm^{-1} indicated were also noticed. After 10% lanthana addition with alumina dual supports, bands for bidentate CO$_2$-chemisorbed species disappeared and bands for unidentate carbonate appeared at 1,517 cm^{-1}. Alfatish et al. had suggested that basic La$_2$O$_3$ interacted with acidic CO$_2$ and formed unidentate carbonate. Further, lanthana addition at 15 wt% showed bands for both unidentate as well as bidentate carbonate species.

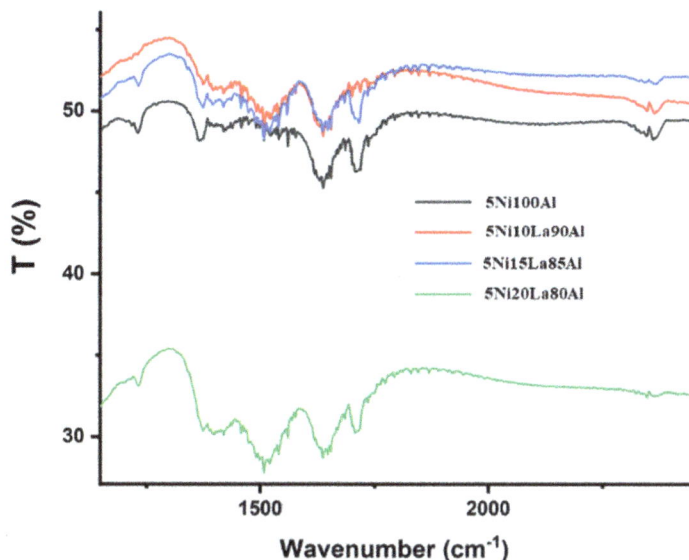

Figure 3.8: The infrared spectra of 5Ni100Al, 5NixLa(100 − x)Al (x = 10,15, 20) samples. With permission from Elsevier.

Example 4: Khatri et al. had synthesized ceria-promoted phosphate–zirconia-supported nickel by wet impregnation of nickel nitrate precursor and then after coimpregnation of ceria nitrate precursor over "8 wt% phosphate + 92 wt% zirconia" support [4]. The catalyst was abbreviated as 10NixCePZr (x = 1, 1.5, 2, 3, 5). The IR spectra of ZrO_2, PZr, 3Ce/PZr, 10NiPZr, 10NiPZr, 10NixCePZr (x = 1, 1.5, 2, 3, 5) catalyst samples are shown in Figure 3.9. The IR spectroscopy of phosphate–zirconia-supported catalyst system had signals of stretching vibration of hydroxyl at 3,426 cm^{-1}, bending vibration at 1,634 cm^{-1}, symmetrical stretching vibrations of P–O (in PO_4^{-3}) at 1,053 cm^{-1}. It has peaks for physically adsorbed CO_2 at 2,351 cm^{-1}, for bidentate carbonate at 1,270 cm^{-1}, for unidentate carbonate at 1380 cm^{-1}, for carboxylate-type CO_2 chemisorbed species at 1,417 cm^{-1}, for stretching vibration of C–H (of bicoordinated format species) at 2,850 cm^{-1} and for the combination of asymmetric stretching of COO and bending vibration of C–H bond at 2,925 cm^{-1}. Overall, it had enriched CO_2 adsorbed species on the catalyst surface. It is noticeable that catalyst samples having both nickel and ceria have diffused to an intense range of peaks from 2,850 to 3,000 cm^{-1}.

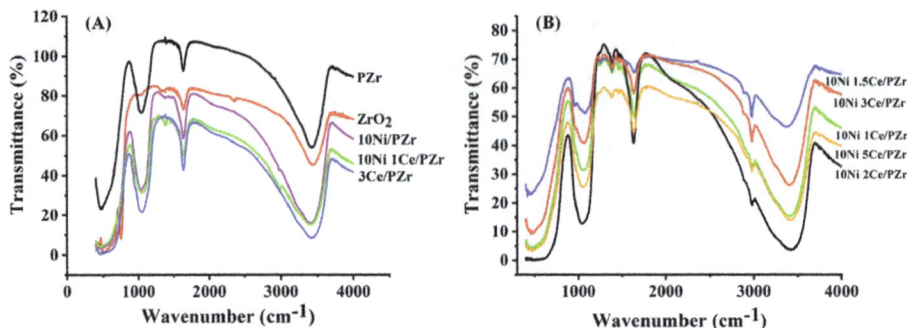

Figure 3.9: The infrared spectra of ZrO_2, PZr, 3Ce/PZr, 10NiPZr, 10NiPZr, 10NixCePZr ($x = 1, 1.5, 2, 3, 5$) samples. With permission from Wiley.

Example 5: Highly ordered mesoporous niobia–zirconia composites were prepared via an evaporation-induced self-assembly (EISA) method by using the following composition: $NbCl_5$:$ZrOCl_2$ 8H_2O:pluronic P123:EtOH = (0.5–4.5):(4.5–0.5):0.08:170 [5]. The catalyst was abbreviated as NbO_2–ZrO_2 (Nb/Zr = 100–x/x; $x = 10$–90). The IR spectra of NbO_2, NbO_2–ZrO_2 (Nb/Zr = 100 – x/x; $x = 10$–90) and ZrO_2 catalyst samples are shown in Figure 3.10. The IR results of pure mesoporous niobia indicate the presence of polymeric (-Nb-O-Nb-)$_n$ species at 880–925 cm^{-1}. On doping with zirconia, this band has decreased sharply. On incorporating 10% zirconium with 90% niobia, IR band at around 3,400 cm^{-1} and 1,630 cm^{-1} band due to stretching and bending of water has been increased, and a new band in the region of 3,710–3,974 cm^{-1} for Nb-OH-Zr has been appeared. This observation was explained by the Kung's model [6]. Replacement of higher valent niobium cation by zirconium cation might leave one excess oxygen anion at each doping center which forms niobinol (Nb-OH) or Nb–OH–Zr species in acidic medium or water.

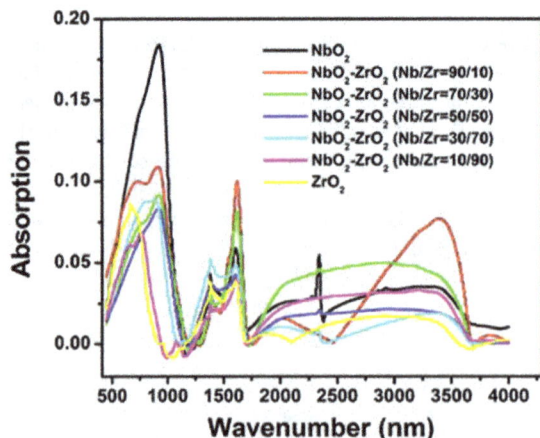

Figure 3.10: The infrared spectra of NbO_2, NbO_2–ZrO_2 (Nb/Zr = 100 – x/x; $x = 10$–90) and ZrO_2 samples. With permission from Elsevier.

References

[1] Al-Fatesh, A. S., Chaudhary, M. L., Fakeeha, A. H., Ibrahim, A. A., Al-Mubaddel, F., Kasim, S. O., Albaqmaa, Y. A., Bagabas, A. A., Patel, R., Kumar, R. *Processes* **2021**, *9*(1), 1–15.

[2] Patel, R., Fakeeha, A. H., Kasim, S. O., Sofiu, M. L., Ibrahim, A. A., Abasaeed, A. E., Kumar, R., Al-Fatesh, A. S. *Mol. Catal.* **2021**, *510*(April), 111676.

[3] Al-mubaddel, F. S., Kumar, R., Lanre, M., Frusteri, F., Aidid, A., Kumar, V., Olajide, S., Hamza, A., Elhag, A., Osman, A. I., Al-Fatesh, A. S. *Int. J. Hydrogen Energy* **2021**, *46*(27), 14225–14235.

[4] Khatri, J., Fakeeha, A. H., Kasim, S. O., Lanre, M. S., Abasaeed, A. E., Ibrahim, A. A., Kumar, R., Al-Fatesh, A. S. *Int. J. Energy Res.* **2021**, No. June, 1–14.

[5] Kumar, R., Ponnada, S., Enjamuri, N., Pandey, J. K., Chowdhury, B. *Catal. Commun.* **2016**, *77*, 42–46.

[6] Kung, H. H. *J. Solid State Chem.* **1984**, *52*(2), 191–196.

4 UV–vis spectroscopy

4.1 Background

When a compound/atom is illuminated with electromagnetic radiation of ultraviolet–visible (UV–visible) range, some wavelength region from the spectrum is absorbed and the rest is transmitted (Figure 4.1A). The absorbed radiation brings an electron excitation into higher energy state according to the rules of odd–even parity (gerade/ungerade[a]) and spin (Figure 4.1B). These rules are selection rules.[b] The transmitted radiation glances over the photoelectric cell and generates the pulsating or alternating current. Alternating current is further amplified (by amplifier) and recorded as the intensity of transmitted radiation (I). Those electronic transitions which cause a "change in parity and without a change in spin" as transition in bonding to nonbonding molecular orbital[c] and charge transition from a metal to a ligand and charge transition from a ligand to a metal[d] are highly intense. Those electronic transitions which cause no change in parity and spin are less intense bands, for example, d–d transition band[e]. Lastly, electronic transition causes no change in parity but changes in spin are very faint.

When the intensity of transmitted radiation is compared with the intensity of incident radiation, transmittance (T) and absorbance (A) are defined. Simply, the measure of the amount of radiation absorbed by a compound is called absorbance (A). Absorbance is equal to the logarithmic ratio of the intensity of incident radiation (I_o) and the intensity of transmitted radiation (I). From the molecular point of view of sample (taken for analysis), absorbance is correlated proportionally to the thickness of medium (b) as well as the molar concentration of solution (c) by Lambert–Beer[f]. The UV spectrum usually plotted absorbance (A) against wavelength (λ) abscissa. Interconversion of intensity data from absorbance (A) to transmittance (T) can be carried out by the formula $A = 2 - \log(T\%)$.[g]

In the case of d^1 or d^6 complex, along with the main peak, some shoulder or split peaks are observed due to asymmetrical electronic distribution, and this effect is explained by the Jahn–Teller effect.[h] It is a very important technique to establish the trend of energy gap in different complexes[i] and to explain the color of a substance.[j] By the help of UV results, a plot of energy state (of d^n system) versus ligand field strength (Dq) is plotted by Orgel, which gives a quantitative idea of energy states of complex (**Annexure I–III, V**). By the help of UV result and Orgel diagram, interelectronic repulsion term (Racha parameter B, which is causing the lowering of energy of lower energy state and increasing the energy of higher energy state) and energy gap between different transitions can be calculated (**Annexure IV**). Further, the energy gap between the plot of E/B versus Dq/B can be plotted (known as Tunabe–Sugano diagrams) which is valid for both weak-field ligand as well as high-field ligand for both the spin-allowed and the spin nonallowed cases. By the help of UV plot, the Orgel diagram and Tanabe–Sugano diagram, correct guess of Racha parameter and Dq can be calculated

https://doi.org/10.1515/9783110656480-004

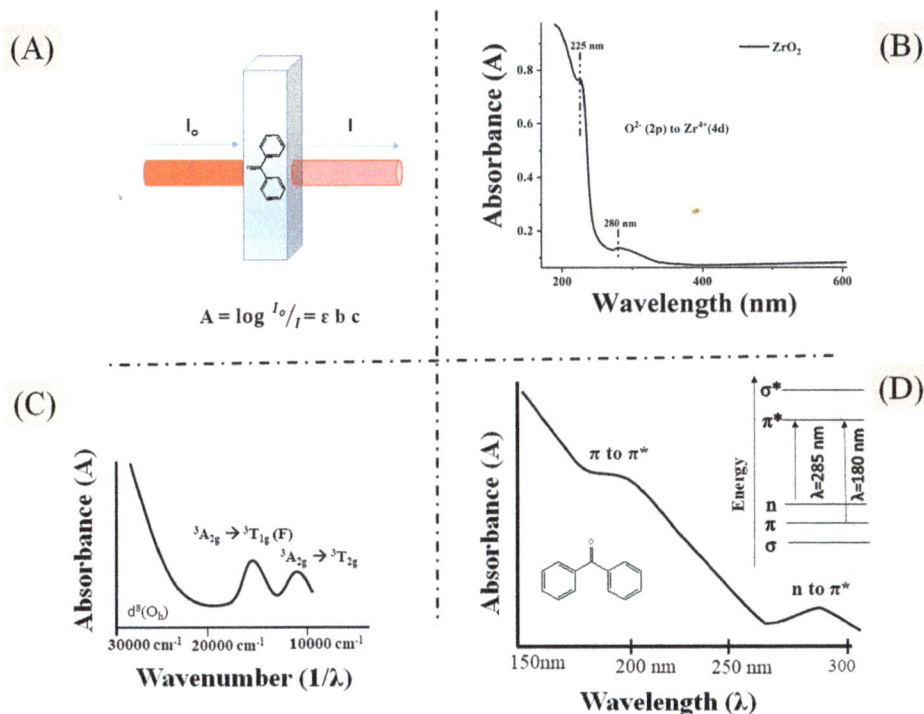

Figure 4.1: (A) Absorption (by samples) and subsequent transmission of UV–vis spectrum across the sample, (B) absorption followed by electron excitation as charge transfer, (C) absorption followed by electron excitation as d–d transition and (D) absorption followed by electron excitation as n to π^* and π to π^* transition.

(**Annexure VI**). If the sample is powder (as in our case) or the particle size of sample is not so small or particle size is comparable to the wavelength of incident light, the relative change in the amount of reflected light off of a source becomes important in place of transmitted light. For such substance, the UV–vis spectroscopy is termed as diffuse reflectance UV–vis spectroscopy. This technique is also useful to calculate the band gap between valence band and conduction band.[k]

4.2 Instrumentation and working principle

A systematic presentation of UV–vis spectrometer is shown in Figure 4.2. Tungsten incandescent filament lamp is used as radiation source for visible region (~350–800 nm), whereas deuterium discharge lamp is used as radiation source for UV region (~175–400 nm). Radiation from a source is allowed to pass through the collimating lens for producing parallel beam and ruled grating (a series of microscopic

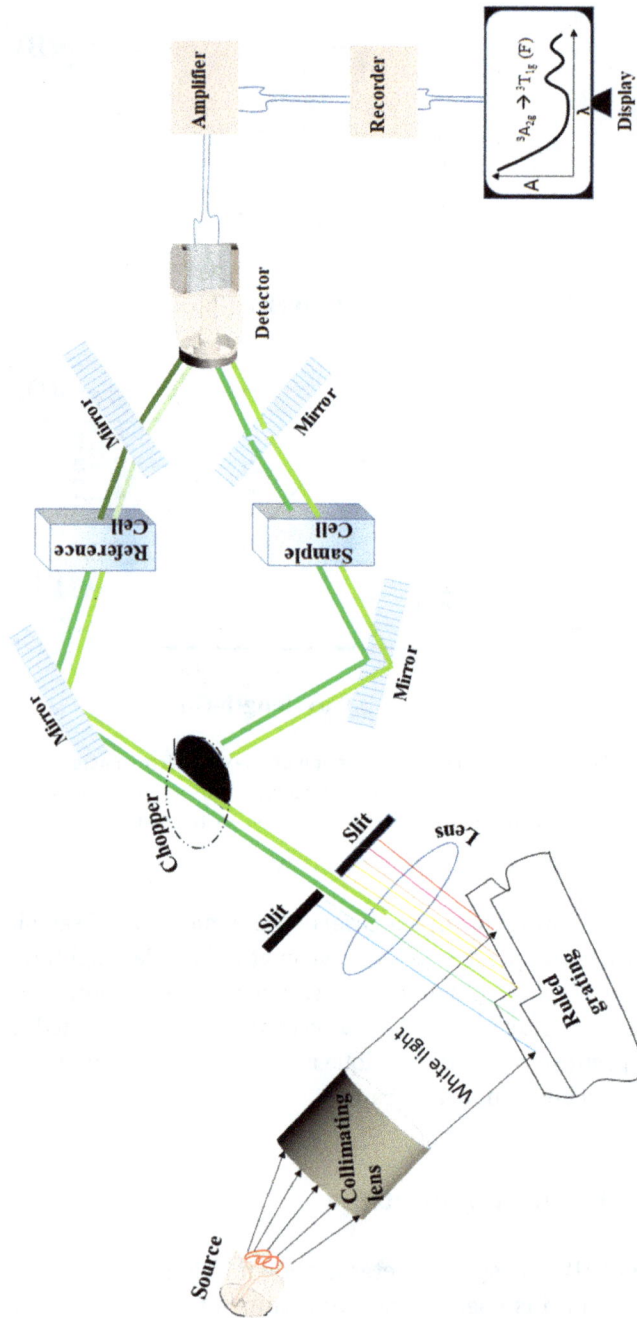

Figure 4.2: A systematic presentation of UV–vis spectrometer.

close parallel triangular grooves engraved onto a reflective surface: 4,000 grooves/mm) for generating specific wavelength or monochromatic beam and focusing lens for focusing beam on the focal plane. Light reflected through grating is diffracted, interfere constructively and generate specific wavelength beam. Further, the monochromatic narrow beam encounters a rotating mirror (chopper) which permits the beam to pass through the sample during half of its period of rotation and reflects the beam to the fixed mirror to the reference cell (containing the reference material/solvent) in the next half period of rotation. The transmitted beams from sample as well as from reference reach to the detector (photo multiplier tube) which converts radiation energy into electrical energy. The signal for absorption due to reference is automatically subtracted and now the net signal is only due to absorption by the sample. The output signal is amplified by an external amplifier and then the recorded absorbance/transmittance versus wavelength. In the same way, the sample is radiated with the whole range of wavelength (~175–800 nm).

When a sample is solid (as in our case), the relative change in the amount of reflected light off of a source becomes important in place of transmitted light. Such spectroscopy is termed as diffuse reflectance UV–vis spectroscopy.

4.3 Glossary

[a]**Gerade (g) and ungerade (u):** The atomic orbital, wave function equation and symmetry are shown in Figure 4.3. Electrons move in different orbitals in different orbits. Electron is moves in "s" orbital in the first orbit. It has angular wave function $(\Theta \, \Phi) = [1/4\pi]^{1/2}$. So, magnitude of wavefunction is constant, whereas the sign of wave function does not change. For moving from one point to the opposite point on wave function, the sign of wavefunction does not change. "s" orbital pursues gerade symmetry. It is shown by "g."

Electron moves in "s" orbital and in "p" orbital in the second orbit. "s" orbital has gerade symmetry (already discussed). "p" orbital has angular wave function $(\Theta \, \Phi) = [3/4\pi]^{1/2} \cos \theta$. For $90° < \theta < 270°$, $\cos \theta$ is negative and so the sign of wavefunction changes in these regions. For moving from one point to the opposite point on "p" wave function, the sign of wavefunction changes. "p" orbital pursues ungerade symmetry. It is shown by "u."

Electron moves in "s," "p" and "d" orbitals in the third orbit. "s" and "p" orbitals have gerade and ungerade symmetry, respectively (already discussed). "d" orbital has angular wave function $(\Theta \, \Phi) = [5/16\pi]^{1/2} (3 \cos^2\theta - 1)$. That means, all negative values of $\cos \theta$ become positive on squaring it $(\cos^2\theta)$. So, the sign of wave function does not change. For moving from one point to the opposite point on "d" wave function, the sign of wavefunction does not change. "d" orbital pursues the gerade symmetry. It is shown by "g."

The molecular orbital and symmetry along with their energy profile are shown in Figure 4.4. In the case of molecular orbital, σ or π bond has overlapping electron cloud

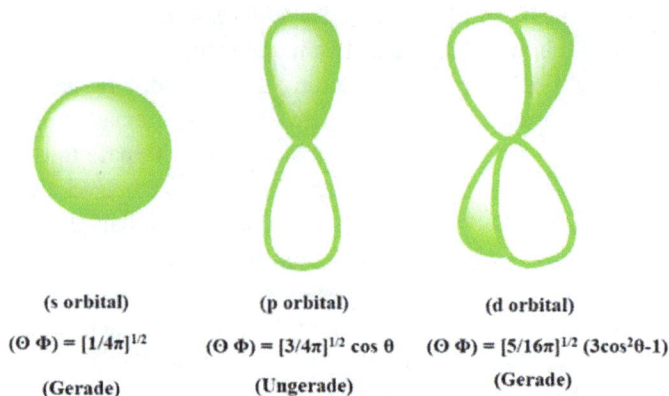

(s orbital)	(p orbital)	(d orbital)
$(\Theta\,\Phi) = [1/4\pi]^{1/2}$	$(\Theta\,\Phi) = [3/4\pi]^{1/2}\cos\theta$	$(\Theta\,\Phi) = [5/16\pi]^{1/2}\,(3\cos^2\theta{-}1)$
(Gerade)	(Ungerade)	(Gerade)

Figure 4.3: The atomic orbital, wave function equation and symmetry.

in the internuclear region. It causes strong attraction of electrons to the nucleus as well as minimum interelectronic and internuclear repulsions. Overall, attraction is umpire over repulsion and it creates bonding between two atoms. σ has the center of symmetry (gerade symmetry) and π has center of asymmetry (ungerade symmetry). If σ^* or π^* has electrons clout out of internuclear region. It causes prevail of interelectronic and internuclear repulsions over electron–nuclear attraction. It destabilizes the bond and called antibonding. σ^* has ungerade symmetry, whereas π^* has gerade symmetry.

[b]**Selection rules:** Selection rule is based on orbital parity and spin. It can be classified into three broad segments. (1) Laporte-allowed and spin-allowed transition: It belongs to such transition which brings "change in parity" and "not change in spin." (2) Laporte-forbidden and spin-allowed transition: It belongs to such transition which brings "no change in parity" and "not change in spin." (3) Laporte-forbidden and spin-forbidden transition: It belongs to such transition which brings "no change in parity" and "change in spin."

I Laporte-allowed and spin-allowed transitions

On irradiation of UV light over an atom, electron transits from one energy level to another by the change of parity (Laporte allowed; g→u, u→g) without changing its spin (spin allowed; $\Delta s = 0$). In other words, it can be said that electron excites from "gerade (g) to next ungerade (u)" or "ungerade (u) to next gerade (u)" orbital wavefunction.[2] This is the allowed transition and known as Laporte-allowed transition. On changing parity, angular quantum number (subsidiary quantum number) of an electron also changes by one unit. So, it can be said that transition which involves change in subsidiary quantum number by 1 is Laporte allowed ($\Delta l = \pm 1$). Due to the fully allowed

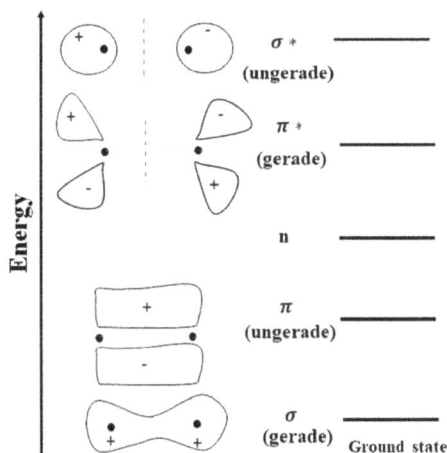

Figure 4.4: The molecular orbital and symmetry along with their energy profile.

transition, the absorption bands are highly intense with molar absorption coefficient (ε) 5,000–10,000 L/mol cm. Charge transfer spectra and bonding to antibonding electron transition are Laporte-allowed and spin-allowed transition, for example, $[TiCl_6]^{2-}$.

II Laporte-forbidden and spin-allowed transitions

Laporte-forbidden and relaxation in Laporte-forbidden electronic transition is pictured in Figure 4.5. Electron transition occurs among d orbital lobes, without change in parity (g→g, u→u) (or without change in subsidiary quantum number; $\Delta l = 0$) and without change in spin. For Laporte-allowed transition, parity should be changed (discussed earlier). But here parity does not change. So, this transition is Laporte forbidden but spin allowed. However, during vibration, some mixing of d and p orbitals occurs which causes partial loss of parity. So, "partial Laporte-allowed transition" is possible. Its absorption band is very faint with molar absorption coefficient (ε) 8–10 L/mol cm, that is, $\left[Ti(H_2O)_6\right]^{3+}$, $\left[V(H_2O)^6\right]^{3+}$.

Tetrahedral complex as well as asymmetric-substituted octahedral (also square planner complex) losses its parity. Relaxation in Laporte-forbidden condition allows frequent mixing of adjacent d orbital and p orbital (d–p mixing) easily, that is, $\left[Co(NH_3)_5Cl\right]^{2+}$, $[MnBr_4]^{2-}$. It causes "spin-allowed" and "Laporte-forbidden relaxed" transitions. Absorption band in asymmetric octahedral complex is more intense than the Laporte-forbidden transition with absorption coefficient (ε) 100–1,000 L/mol cm, that is, $\left[Co(NH_3)_5Cl\right]^{2+}$. Its absorption band in the tetrahedral complex is more intense than the Laporte-forbidden transition with absorption coefficient (ε) 10–100 L/mol cm, that is, $[PdCl_4]^{2-}$.

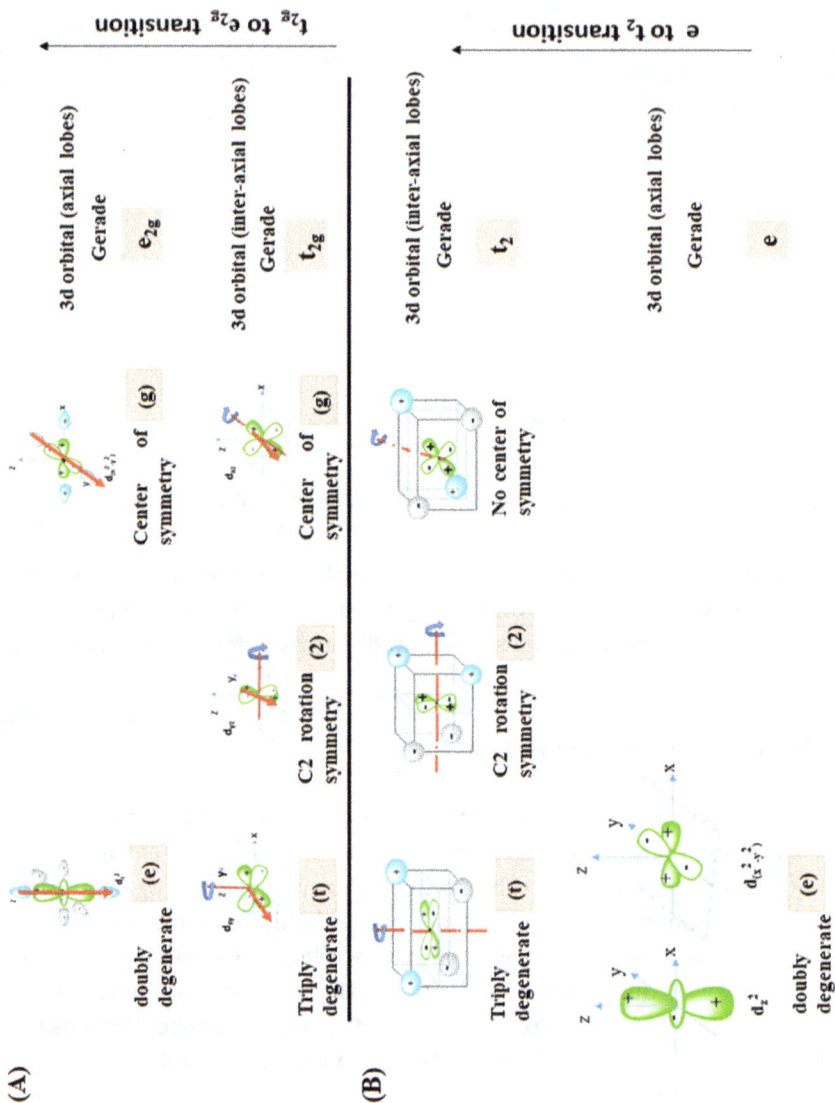

Figure 4.5: (A) Laporte-forbidden and spin-allowed transition: t_{2g} to e_g transition. (B) Relaxation in Laporte-forbidden and spin-allowed transition: e to t_2 transition.

III Laporte-forbidden and spin-forbidden transitions

The electronic configuration modification during Laporte-forbidden and spin-forbidden electronic transition is shown in Figure 4.6. d^5 low-spin octahedral complexes have "all half-filled d orbital," that is, $\left[Mn(H_2O)_6\right]^{2+}$. So, now any d–d transition causes spin reversal of electrons (as two electrons in the same orbital component should align opposite or should have the opposite spin). Overall, it can be said that electron transition between d orbital lobes (low spin complex) causes no change in parity (g→g; $\Delta l = 0$) and change in electron spin ($\Delta s \neq 0$). That means it is both Laporte forbidden and spin forbidden. The absorption band of such compounds is very faint with absorption coefficient (ε) 10^{-3}–1 L/mol cm, that is, $\left[Mn(H_2O)_6\right]^{2+}$.

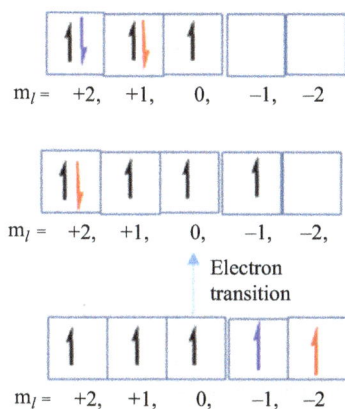

Figure 4.6: The electronic configuration modification during Laporte-forbidden and spin-forbidden electronic transition.

cCharge transfer spectra: Charge transfer spectra are Laporte-allowed and spin-allowed transition. Due to fully allowed transition, the absorption bands are highly intense with high molar absorption coefficient (ε). Charge transfer takes place chiefly in two ways:

A Ligand-to-metal charge transfer

Charge is transferred from the "filled p orbital of ligand orbital" to "vacant d orbital of metal." If a metal has high oxidation state, then there is strong effective nuclear charge at the vacant d orbital of metal which can accommodate the ligand charge, that is, VO_4^{3-}. As in VO_4^{3-}, V^{5+} ion is surrounded tetrahedrally by four O^{2-}. V^{5+} has vacant 3d orbital. O^{2-} has an electron in the p orbital. High oxidation state (+5) creates strong effective nuclear charge at its vacant d orbital region which holds four O^{2-} tetrahedrally.

As we move from VO_4^{3-} to CrO_4^{2-} to MnO_4^-, the oxidation state of metal increases (+5 to +6 to +7), respectively. As the oxidation state of metal increases, effective nuclear charge on vacant "d" orbitals becomes stronger which holds the charges of oxide anion easily. So, energy required for electron transfer from O^{2-} to the metal ion decreases in the order. As we move from VO_4^{3-} to NbO_4^{3-} to TaO_4^{3-}, the oxidation state of metal remains constant but the distance between nucleus and vacant "d" orbitals increases. So, effective nuclear charge on the distant vacant d orbitals becomes worse. So, it does not hold O^{2-} easily. So, the energy required for electron transfer from O^{2-} to the metal ion increases in the order. In the case of 4d and 5d metal series, the energy requirement for electron transfer is very high (in the UV region). So, it looks colorless. The ligand anion with low electron affinity transfers charge easily to a metal center. If Cl^- ligand is replaced by Br^- or I^-, more loosely bound ions (low affinity electrons) are available for transfer to the metal center. So, energy required for electron transfer from O^{2-} to the metal ion decreases in the order. The order of energy required for electron transfer from O^{2-} to the metal ion is presented in Figure 4.7.

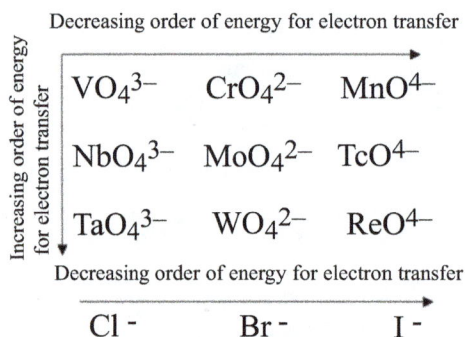

Figure 4.7: The order of energy required for electron transfer from O^{2-} to the metal ion.

B Metal-to-Ligand charge transfer

Charge is transferred from the filled d orbital of the metal to the vacant antibonding orbital (π^*) of ligand as CN^- (Figure 4.8). It gives metal-to-ligand charge transfer spectra in the visible range. So, these compounds are colored.

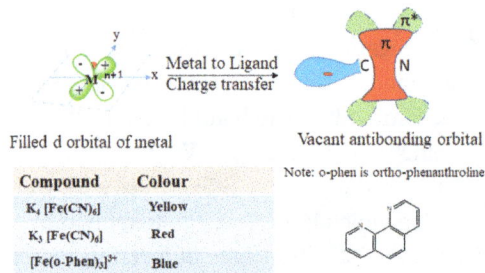

Figure 4.8: Charge transfer from the filled d orbital of metal to the vacant antibonding orbital (π^*) of ligand and the color of compound.

[d]Electronic transition from bonding molecular orbital to the antibonding molecular orbital: During transition from bonding molecular orbital to the antibonding molecular orbital ($\sigma \to \sigma^*$; $\pi \to \pi^*$), parity of electron changes (g→u; u→g) without change in spin. This transition is Laporte allowed and spin allowed ($\Delta s = 0$) (Figure 4.9). Among $\sigma \to \sigma^*$ and $\pi \to \pi^*$; later one needs less energy for electron transition because $\pi \to \pi^*$ has less energy gap than $\sigma \to \sigma^*$. So, it is very common in the unsaturated organic system. Electronic transition from the nonbonding electron to the antibonding π orbital ($n \to \pi^*$) is symmetric forbidden but energy-wise favorable (because it has least energy gap). So, under vibronic coupling, nonbonding orbital mixes with π^* orbital and electron transition from $n \to \pi^*$ will be possible in low intensity at about 280 nm UV radiation. In photochemistry, such transition is very familiar with carbonyl compounds.

$\sigma \to \sigma^*$ electronic transition is possible in alkane around 150 nm UV radiation source, whereas $\pi \to \pi^*$ transition is initiated around 170–190 nm absorption band. $\pi \to \pi^*$ transition is very frequent in organic compounds. In conjugated diene, π electrons are delocalized over the organic skeleton and it generates an equivalent π ring bonding molecular orbital and equivalent π^* ring antibonding molecular orbital having less energy gaps than the isolated ones. So now $\pi \to \pi^*$ transition is possible with longer wavelength (less energy) UV radiation (220 nm). With increasing conjugation of double bond, transition wavelength may be progressed to 600 nm.

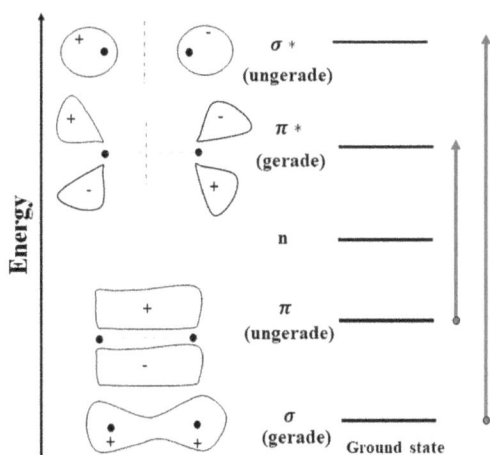

Figure 4.9: Laporte-allowed and spin-allowed ($\Delta s = 0$) transition among molecular orbitals.

[e]**d–d transfer spectra:** Electrons revolve around the nucleus with some angular momentum. As these are vector quantities, the resultant angular momentum constitutes various quantified energy states between 0 and the "sum of total angular momentum" (represented by L). Again, these electrons are spinning on their axes with spin momentum. As these are vector quantities, the resultant spin momentum constitutes various quantified energy states between 0 and the "sum of total spin momentum."

These spin states are oriented in 2S + 1 manner (called multiplicity; represented by M) along the magnetic field. The total momentum state is presented as ^{M}L. Further in the unsymmetric ligand field, these states are splitted into various energy states. Electron transition between these energy states is Laporte-forbidden and spin-allowed transitions (transition in same parity with same spin). These spectra are less intense than the charge transfer spectra. The energy states (E) versus ligand field strength (Dq) are presented and known as the Orgel diagram, if the ground state term is taken as the horizontal axis. This horizontal line provides the constant reference point and the other states can be plotted relative to the ground state. Interelectronic repulsion term in metal ion is presented as the Racha parameter (B). The plot of E/B versus Dq/B is known as the Tunabe–Sugano diagrams. Orgel diagram is used for weak-field spin-allowed case, whereas Tunabe–Sugano diagram is used for both weak field and strong field in spin-allowed and spin not-allowed cases.

Here, the oral diagram, spectra for d^1–d^9 and d^6–d^4 cases are discussed in **Annexure I**. Spectra for d^2–d^8 and d^3–d^7 system are discussed in **Annexure II**. Spectra for d^5 system are discussed in **Annexure III**. Racha parameter (B) calculation is discussed in **Annexure IV**. Tanabe–Sugano diagrams for d^6 complex is discussed in **Annexure IV**. Guess of B and Dq is discussed in **Annexure VI**.

^{f}**Lambert–Beer's law:** Absorption (by samples) and subsequent transmission of UV–vis spectrum across the sample is shown in Figure 4.10. Lambert states that when the electromagnetic wave passes through a sample, the rate of decrease of transmitted radiation is proportional to the increase in the thickness of the medium (db) as well as the intensity of the incident light (I_0).

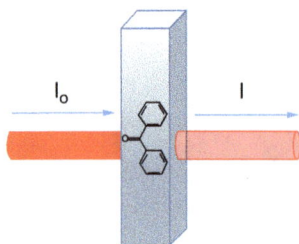

Figure 4.10: Absorption (by samples) and subsequent transmission of UV–vis spectrum across the sample.

Form the Lambert statement,

$$-dI_0 = kI_0 db$$

$$-\frac{dI_0}{db} = kI_0$$

$$I = I_0 10^{-k'b}$$

$$\log\frac{I_0}{I} = k'b$$

In terms of absorbance, $\log \dfrac{I_o}{I} = A \propto b$

Beer states that when electromagnetic wave passes through a sample, the rate of decrease of transmitted radiation is proportional to the increase in the concentration of the medium (dc) as well as the intensity of the incident light (I_o):

$$-dI_o = kI_o dc$$

$$-\frac{dI_o}{dc} = k\, I_o$$

$$I = I_o\, 10^{-k''c}$$

In terms of absorbance, $\log \dfrac{I_o}{I} = A \propto c$

The combined statement of Lambert–Beer can be stated that when an electromagnetic wave passes through a sample, absorption is proportional to the thickness of the medium (b) and concentration of the medium (c):

$$A \propto bc$$

$$A = \varepsilon\, b\, c$$

where ε is the molar absorption coefficient or molar absorptivity.

[g]**Interconversion of intensity data from absorbance (A) to transmittance (T):**
The absorbance (A) is studied by monitoring both incident intensity (I_o) and transmitted intensity of radiation (I_o) as $A = \log(I_o/I)$. Transmittance (T) is defined by $T = I/I_o$. So, percentage transmittance can be shown as $T\% = (I/I_o).100$. Putting log on both sides, this equation will be $\log(T\%) = \log\left(\dfrac{I}{I_o}\right) + \log 100$

$$\log(T\%) = -\log\left(\frac{I_o}{I}\right) + 2$$

After putting the absorbance term, $\log(T\%) = -A + 2$

$$A = 2 - \log(T\%)$$

In this way, interconversion of intensity data of either transmittance (T) or absorbance (A) can be carried out by the formula $A = 2 - \log(T\%)$. UV–vis spectrum is presented as "transmittance versus frequency" or "absorbance versus frequency."

[h]**Jahn–Teller distortion:** Axial e_g orbitals are directly pointing toward ligands. Unsymmetrical electronic distribution in axial e_g orbital (of metal orbital) causes noticeable repulsion to the ligands from its extra electron axis. So, the ligand will move further away from that extra excess electron axis causing distortion in shape. It is

known as Jahn–Teller distortion, that is, high-spin d^4 system $(t_{2g}^3 e_g^1)$: Cr^{2+}, Mn^{3+}; low spin d^7 system $(t_{2g}^6 e_g^1)$: Co^{2+}, Ni^{3+}; d^9 system $(t_{2g}^6 e_g^3)$: Cu^{2+} as $Cu(H_2O)_6]^{2+}$ $[Cu(en)_2H_2O)_2]^{2+}$, $[Cu(en)_3]^{2+}$.

As for d^9 system ($t_{2g}^6 e_g^3$): If e_g^3 is configured like $d^2_{(x^2-y^2)}$, $d^1_{(z^2)}$, then four ligands along extra electron axis (x and y axes) are repelled to the greater extent and these ligands move farther away. It is also called "z-in case". If e_g^3 is configured like $d^1_{(x^2-y^2, d^2(z^2))}$, then two ligands along extra electron axis are repelled to the greater extent and these ligands move farther away. It is also known as z-out case. Energetically, two bond elongations (along z-axis) are more favorable than four bond elongations (along x and y axes) (Figure 4.11). So, $d^1_{(x^2-y^2), d^2(z^2)}$ configuration is more favorable. It presented distorted octahedral shape.

Consequences 1. Six-coordinated $[Cu(H_2O)_6]^{2+}$ complex possesses distorted octahedral structure. Two of the H_2O (ligand) in "z" direction are distant than four ligands on the "xy" plane. Elongated z-axis ligands (H_2O) are less tightly attached to the metal center and so exchanged easily in a reaction.

Consequences 2. The general trend of chelation is that "greater the number of chelation greater will be the stability." But due to John–Teller distortion, the reverse trend is seen in $[Cu(en)_3]^{2+}$, $[Cu(en)_2H_2O)_2]^{2+}$ complexes. $[Cu(en)_3]^{2+}$ has three chelate rings in which two are under strain due to elongated z-direction bond. The strain in $[Cu(en)_3]^{2+}$ makes it less stable than $[Cu(en)_2H_2O)_2]^{2+}$.

For d^1 complex, electron excites from the ground state (t_{2g}) to the excited state (e_g) in two ways. It may enter into $d_{x^2-y^2}$ or d_{z^2}. Due to the short life time of different excited states, it does not permit the attainment of stable equilibrium configuration of the complex. Unsymmetrical electronic distribution in axial e_g orbital (axial component of metal d orbital: $d_{(x^2-y^2)}$, $d_{(z^2)}$) leads splitting of excited states into a main and a shoulder peak. So, UV spectra of d^1 complex shows a shoulder peak along with a single peak due to the Jahn–Teller distortion (Figure 4.12). Again, for d^6 complex; electron excites from the ground state (t_{2g}) to the excited state (e_g) in two ways. It may enter into $d_{x^2-y^2}$ or d_{z^2}. Unsymmetrical electronic distribution in axial e_g orbital (axial component of metal d orbital: $d_{(x^2-y^2)}$, $d_{(z^2)}$) leads splitting of excited states into two clear peaks (Figure 4.13). So, UV spectra of d^6 complex shows clear two peaks due to the Jahn–Teller distortion.

[i]**Factor affecting the magnitude of Δ:** Magnitude of Δ increases with (1) higher order ligand geometry around the central metal as VCl_4 has tetrahedral geometry, $\Delta = 7{,}900$ cm^{-1}; $[VCl_6]^{2-}$ has octahedral geometry; $\Delta = 15{,}400$ cm^{-1}. (2) Higher orbit "d" orbital of central metal $[Fe(H_2O)_6]^{3+}$ has 3d orbital, $\Delta = 14{,}000$ cm^{-1}; $[Ru(H_2O)_6]^{3+}$ has higher 4d orbital, $\Delta = 28{,}600$ cm^{-1}. (3) Higher oxidation state of central metal $[Ru(H_2O)_6]^{2+}$ has +2 oxidation state, $\Delta = 19{,}800$ cm^{-1}; $[Ru(H_2O)_6]^{3+}$ has +3 oxidation state, $\Delta = 28{,}600$ cm^{-1}.

Magnitude of Δ increases with higher field strength of ligand. The order of increasing field strength is based on experimentally determined series. However, increasing

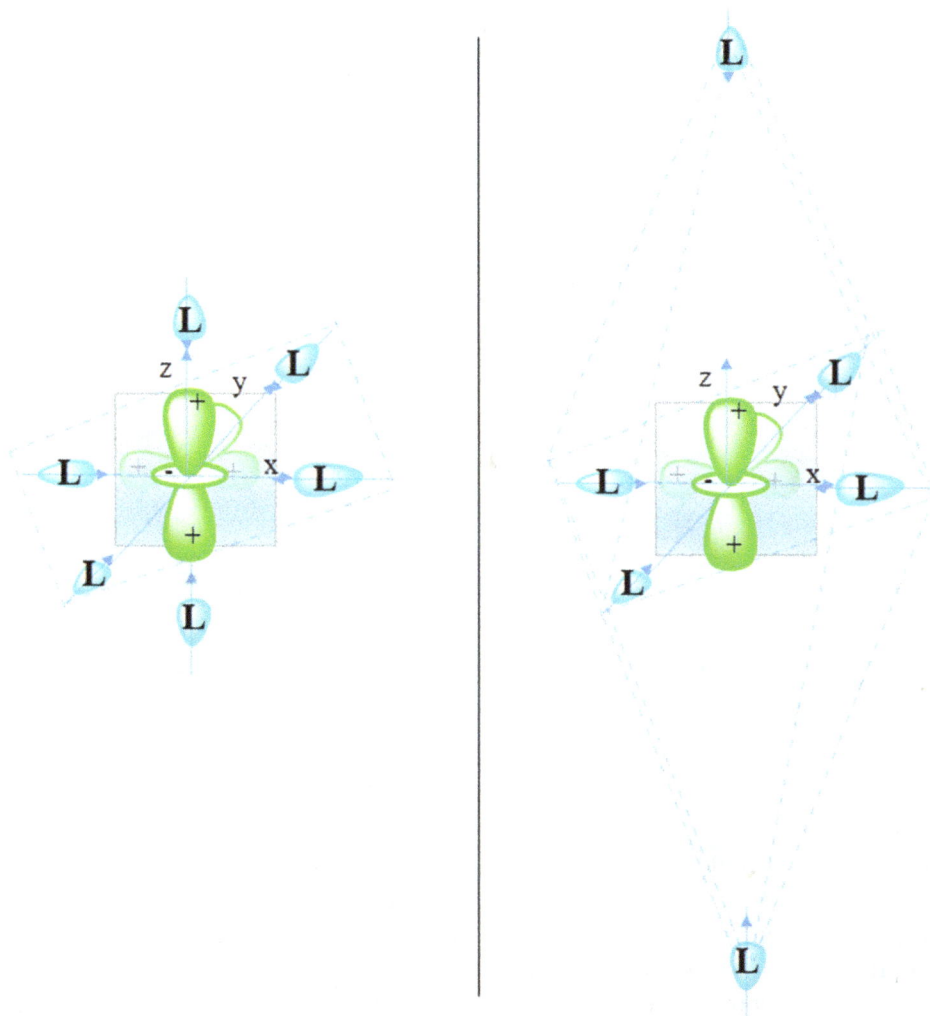

Figure 4.11: Energy-favorable two bond elongations (along *z*-axis) than the rest four bonds (along *x* and *y* axes) under Jahn–Teller distortion.

filed strength can be hardly correlated with increasing "σ" donation (halide $< O < N < C$) and increasing "π" acceptance from ligand ($CN^- < CO$). "π" donation from ligand to metal competes with σ donation capacity of ligand or overall decrease in the σ donation capacity of ligand. So, field strength of OH^- ("σ" donation by lone pair over oxygen and π donation by anion) is less than H_2O (only σ donation by lone pair over oxygen).

Overall the order of increasing field strength of ligand is as follows:

$I^- < Br^- < S_2^{2-} < SCN^- < Cl^- < NO_3^- < F^- < OH^- < EtOH < oxalate < H_2O < NCS^- < NH_3, py. <$ ethylenediamine < dipyridyl < *o*-phenanthroline $< NO_2^- < CH_3^-, C_6H_5^- < CN^- < CO$

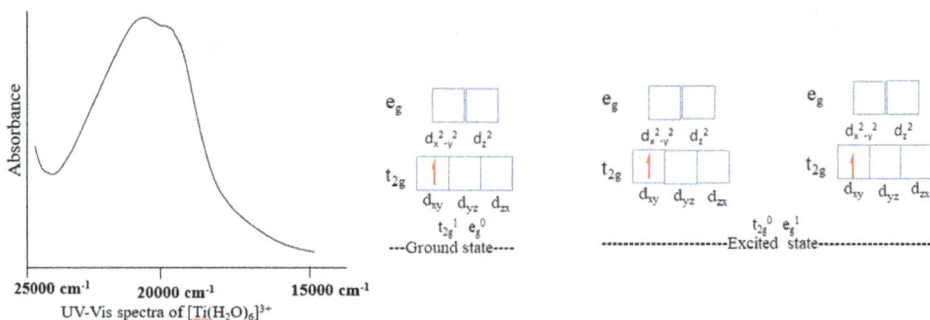

UV-Vis spectra of [Ti(H₂O)₆]³⁺

Figure 4.12: Electronic transition of d¹ complex in ground as well as excited state and UV spectra of d¹ complex.

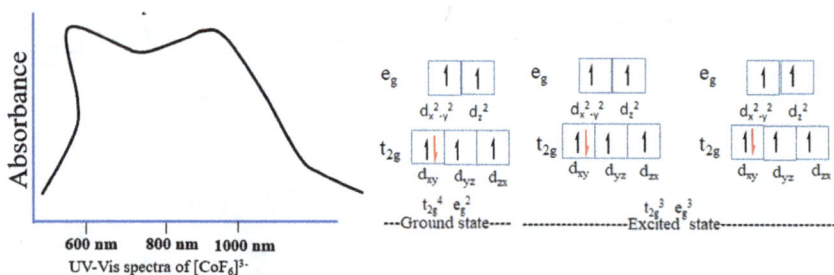

UV-Vis spectra of [CoF₆]³⁻

Figure 4.13: Electronic transition of d¹ complex in ground as well as excited state and UV spectra of d⁶ complex.

ʲ**Color of the complex:** If an electromagnetic wave (belongs to visible region; 400–800 nm) is falling on the metal complexes, electrons absorb peculiar energy band from electromagnetic wave and jump to higher levels. These bands are called absorption bands. No rest electromagnetic band (after abstracting the absorption band) comes out of complexes. These bands are called transition bands. It is a complementary color. It is seen by our eyes. The color band, the corresponding wavelength in nm and frequency (wavenumber) in cm^{-1} are shown in Figure 4.14.

$[Fe(CN)_6]^{3-}$: Strong ligand CN^- causes large splitting or large energy gap. So, blue band is adsorbed from light and electrons excite to the higher energy level. So, it has complementary orange color which will be visible to our eyes. $[Fe(H_2O)_6]^{3+}$: ligand H_2O causes moderate splitting or moderate energy gap. So, green band is adsorbed from light and electrons excite to higher energy level. So, it has complementary red color.

$[V(H_2O)_6]^{3+}$: Strong charge on central meal causes large splitting or large energy gap. So, violet band is adsorbed from light and electrons excite to higher energy level. So, $[V(H_2O)_6]^{3+}$ has complementary yellow color. $[V(H_2O)_6]^{2+}$: Moderate charge on central meal causes moderate splitting or moderate energy gap. So, yellow band

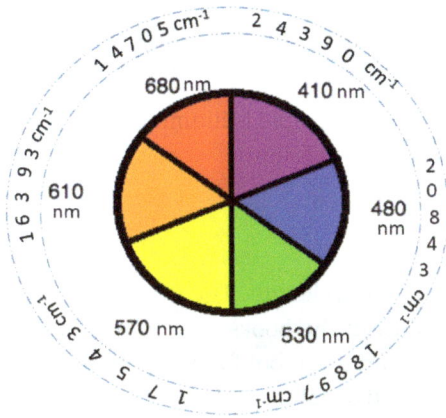

Figure 4.14: The color band, the corresponding wavelength in nm and frequency (wavenumber).

is adsorbed from light and electrons excite to the higher energy level. So, it has complementary violet color.

$[CrF_6]^{3-}$: The ligand field strength is $F < H_2O < NH_3 < CN$. As per the ligand strength, splitting proceeds which create the energy gap. Red band is adsorbed. It signs complementary green color. $[Cr(H_2O)_6]^{3+}$: Yellow band is adsorbed. It signs complementary violet color. $[Cr(NH_3)_6]^{3+}$: Blue band is adsorbed. It signs complementary orange color. $[Cr(CN)_6]^{3-}$: Violet band is adsorbed. It signs complementary yellow color.

[k]**Band gap:** If particle size of sample is comparable to the wavelength of incident light, the behavior of light traveling inside the light scattering specimen is described by Kubelka and Munk model (K-M model) [1]. Light of intensity "i" is travels from un-illuminated surface and intensity "j" travels from the illuminated surface along the path "x" than the change in intensity of light along with path difference:

$$-di = -(S+K)idx + Sjdx$$

$$dj = -(S+K)jdx + Sidx$$

where S is the portion of light scattered per unit vertical length (known as K-M scattering coefficients) and K is the portion of light absorbed per unit vertical length (known as K-M absorption coefficient).

The reflectance of infinitely thick sample [2] (R_∞) is defined as the ratio of reflectance of sample (R_{sample}) and reflectance of standard ($R_{standard}$). $R_\infty = R_{sample}/R_{standard}$. The ratio of K/S is related to the reflectance as follows:

$$\frac{R_{sample}}{R_{standard}} = 1 + \frac{K}{S} - \sqrt{\left(\frac{K^2}{S^2} + \frac{2K}{S}\right)}$$

$$\frac{K}{S} = \frac{(1 - R_\infty)^2}{2R_\infty}$$

When sample scatters in perfectly diffuse manner (or sample is illuminated at 60 °C incidence), the ratio of K/S is related to the band gap as follows:

$$\left(\frac{K}{S} hv\right)^2 = C_1 (hv - E_g)$$

In this case, S becomes constant with respect to wavelength. Then K/S is directly proportional to the absorption coefficient K. Most of the books use "α" symbol for absorption coefficient. After putting S constant and α symbol for absorption coefficient equation, the above equation can be modified as follows [3]:

$$(\alpha hv)^2 = C_2 (hv - E_g)$$

From the UV spectrometer, we have obtained the data of absorbance (A). Dividing absorbance by thickness (t), absorption coefficient can be calculated as $= A/t$. Generally, we are taking 500 nm thickness. The energy term can be calculated by $hv = (hc/\lambda) = 1,240/\lambda$; where λ is the wavelength and unit of energy in eV. On taking plot of $(\alpha hv)^2$ versus hv (Y – axis versus X – axis); C_1 will be the slope and $C_1 E_g$ will be the intercept (of tangent of curve) on x-axis. If you begin y-axis from zero, then intercept (of tangent of curve) on x-axis will be the E_g of sample. A typical plot for calculating band gap is shown in Figure 4.15.

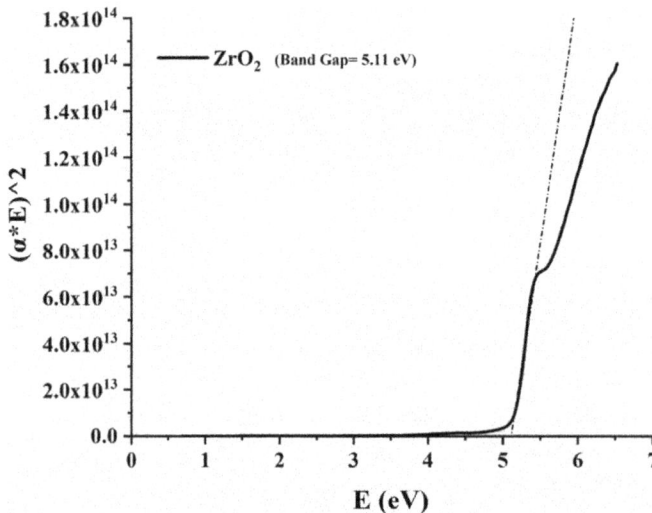

Figure 4.15: A typical plot for calculating band gap.

4.4 Analysis of UV–vis profile

Example 1: SBA-15-supported Bi catalyst was prepared by the sol–gel method under hydrothermal treatment through the following gel composition: Pluronic P123:H_2O:HCl:n-butanol:tetraethyl orthosilicate:Bi$(NO_3)_2$ $6H_2O$ = 0.017:200:5.4:1.325:1:0.01–0.05. The sample is abbreviated as Bi-SBA-15 (Bi/Si = (1–5)/100) [4]. The UV–vis spectra of SBA-15 and Bi-SBA-15 (Bi/Si = (1–5)/100) samples are shown in Figure 4.16. The catalyst showed UV spectra centered about 215 nm attributed to the ligand-to-metal charge transfer spectra, where metal Bi is in tetrahedral coordination in the silica network. For bulk Bi_2O_3, generally a band at 400 nm was observed. But in SBA-15-supported Bi catalyst, 400 nm UV band was absent. It indicates that the bulk Bi_2O_3 does not exist over SBA-15 support, and bismuth is well dispersed over silica frame. As bismuth loading increased from 1 to 5 mol%, the intensity of UV band at 215 nm increased sharply. It indicates that more amount of bismuth is coordinated in the tetrahedral coordination.

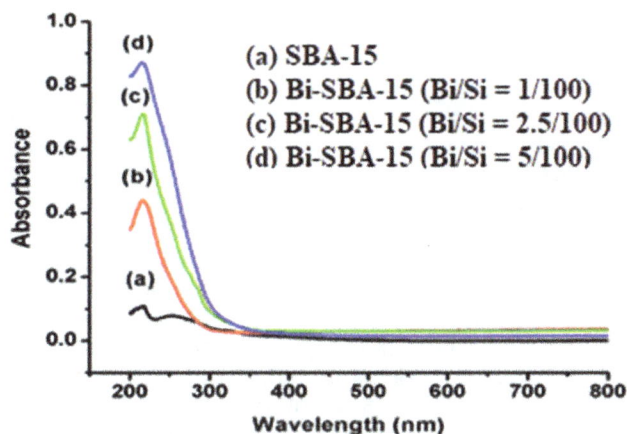

Figure 4.16: The UV–vis spectra of SBA-15 and Bi-SBA-15 (Bi/Si = (1–5)/100) samples. With permission from Elsevier.

Further, SBA-15-supported 1 mol% Bi catalyst was trimethylsilylated by passing methoxytrimethylsilane vapor under argon stream. To confirm silylation, UV–vis–NIR spectra of silylated Bi-SBA-15 (Bi/Si = 1/100) was taken (Figure 4.17). It shows bands after 1,300 nm, which can be assigned to the overtones and combination bands of CH_3 groups. It indicated that the catalyst is silylated properly.

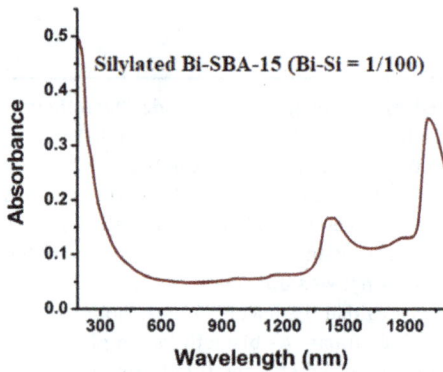

Figure 4.17: UV–vis–NIR spectra of silylated Bi-SBA-15 (Bi/Si = 1/100). With permission from Elsevier.

Example 2: Al-Fatesh et al. [1] prepared 5 wt% Ni dispersed over "γ-alumina doped with 3 wt% MO$_x$ (M = Ti, Si, Mo, W)" by mechanical mixing of nickel nitrate precursor salt with "γ-alumina doped with 3 wt% SiO$_2$" followed by calcination [5]. The catalyst name was abbreviated as 5Ni3MAl (M = Si, W, Ti, Mo). The UV–vis spectra of 5Ni3MAl (M = Si, W, Ti, Mo) samples are shown in Figure 4.18. 5Ni3SiAl and 5Ni3WAl samples show peaks at 250–350 nm for charge transfer from O^{2-} to Ni^{2+} (octahedral coordination), 410 nm for d–d transition from $^3A_{2g}$(F) state to $^3T_{1g}$ (P) state of Ni^{2+} (octahedral coordination) and 593 and 634 nm for d–d transition from $^3T_1$(F) state to $^3T_1$(P) state of Ni^{2+} (tetrahedral coordination). It indicated the presence of Ni^{2+} in both tetrahedral and octahedral coordination (or the presence of NiAl$_2$O$_4$ phase) in SiO$_2$- and WO$_x$-doped samples. Rest samples 5Ni3TiAl and 5Ni3MoAl had no peak for Ni^{2+} in tetrahedral coordination (absence of NiAl$_2$O$_4$ phase). 5Ni3TiAl had additional UV–vis peak at 740 nm for Ni^{2+} in octahedral coordination in the bulk NiO, which means titania-doped system has the presence of free NiO (less interacting NiO or bulk NiO). 5Ni3MoAl showed a hump in the 280–330 nm which is attributed to the charge transfer from O^{2-} to M^{6+} (octahedral/tetrahedral coordination). It indicated the possible presence of NiMoO$_4$ or Al$_2$(MoO$_4$)$_3$ phase.

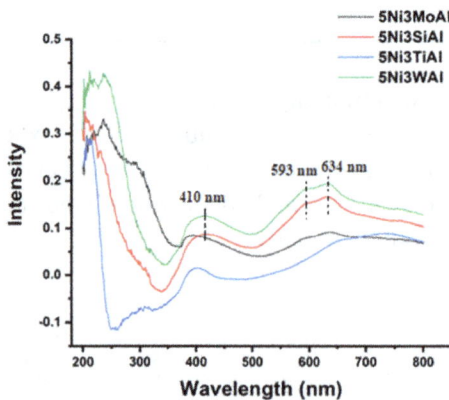

Figure 4.18: The UV–vis spectra of 5Ni3MAl (M = Si, W, Ti, Mo) samples. With permission from MDPI.

Example 3: Tsoncheva et al. had prepared SBA-15-supported Co sample, KIT-6-supported Co sample and KIT-5-supported Co sample by impregnation of 6 wt% cobalt nitrate over the support [6]. SBA-15, KIT-6 and KIT-5 are silicates having two-dimensional hexagonal cylindrical mesopores, three-dimensional two interpenetrating branched networks of cylindrical pores (Ia3d symmetry) and three-dimensional highly interconnected cage-like mesoporous (face-centered cubic structure), respectively. The sample is abbreviated as Co/support; support is SBA-15, KIT-6 or KIT-5. UV spectra of Co/SBA-15, Co/KIT-6 and Co/KIT-5 are shown in Figure 4.19. UV–vis spectra of these samples showed broad peaks at about 400 and 700 nm for Co^{3+} ions in the octahedral environment, whereas wide bands at 538 and 640 nm for d–d band transition $^4A_2(F){\rightarrow}^4T_1(P)$ of Co(II) ions in the tetrahedral coordination.

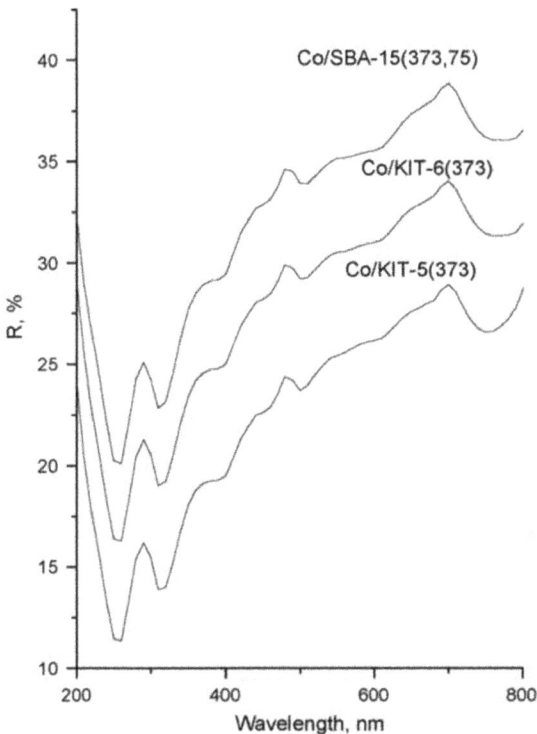

Figure 4.19: UV–vis spectra of Co/SBA-15, Co/KIT-6 and Co/KIT-5. With permission from Elsevier.

Example 4: Khatri et al. prepared ceria-promoted phosphate–zirconia-supported nickel catalyst by the wet impregnation method of nickel nitrate precursor and then after coimpregnation of ceria nitrate precursor over phosphate–zirconia [7]. The catalyst was abbreviated as 10NixCePZr ($x = 1$, 1.5, 2, 3, 5). The UV–vis spectra of ZrO$_2$, PZr, 3Ce/PZr, 10NiPZr and 10Ni1CePZr catalyst samples are shown in Figure 4.20. The UV spectroscopy of ZrO$_2$ shows the UV band at 226 and 280 nm for the charge transfer from O^{2-} to Zr^{4+}. On addition of phosphate, the charge transfer bands of ZrO$_2$ were demolished. On further addition of ceria showed the UV band at the Ce/PZr catalyst at 280 nm again for the charge transition of O^{2-} → Ce^{3+}/Ce^{4+}. Phosphate–zirconia-supported nickel catalyst (Ni/PZr) shows the band at 286 nm for the charge transfer from O^{2-} → Ni^{2+} (O$_h$), at about 423 nm for the d–d transition from ^{3}A$_2$g(F) → ^{3}T$_1$g(P) state of Ni^{2+} (O$_h$) and at about 730 nm for the d–d transition from 3A$_2$g(F) → 3T$_1$g(F) state of Ni^{2+} (O$_h$). After the addition of ceria promoter in Ni/PZr catalyst (NiCe/PZr), the band at 280 nm is intensified, indicating the charge transfer band of O^{2-} to Ni^{2+} (O$_h$) and O^{2-} → Ce^{3+}/Ce^{4+} (O$_h$). However, the d–d transition band at about 423 nm for ^{3}A$_2$g(F) → ^{3}T$_1$g(P) state of Ni^{2+} (O$_h$) was demolished with the increasing ceria loading. It indicates modulation in the coordination environment about Ni^{2+} due to the presence of ceria.

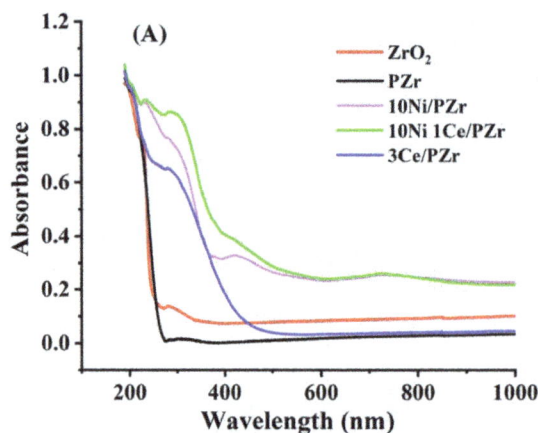

Figure 4.20: The infrared spectra of ZrO$_2$, PZr, 3Ce/PZr, 10NiPZr and 10Ni1CePZr catalysts. With permission from Wiley.

Example 5: Khatri et al. prepared ceria-promoted phosphate–zirconia-supported nickel sample by the above discussed method [7]. The catalyst was abbreviated as 10NixCePZr ($x = 1$, 1.5, 2, 3, 5). The infrared spectra of ZrO$_2$, PZr, 3Ce/PZr, 10NiPZr, 10Ni1CePZr catalyst samples are shown in Figure 4.21. The group studied the band gap between valence band and conduction band of the sample. The band gap of pure zirconia, phosphate zirconia and phosphate–zirconia-supported Ni are 5.05, 4.81 and 3.07 eV, indicating lower band gap on successive incorporation of phosphate and nickel. On further addition of ceria, the band gap rose up to 3.25 eV. However, on successive addition of ceria from 1% to 5%, band gap decreased continuously from 3.25 to 2.89 eV. These observations indicated that the electron transfer from the valence band to the conduction band is easiest in phosphate–zirconia-supported Ni sample among non-ceria samples. Again, ceria addition is found to be beneficial to reduce the band gap further, which pivots the path of easy electron transfer from the valence band to the conduction band.

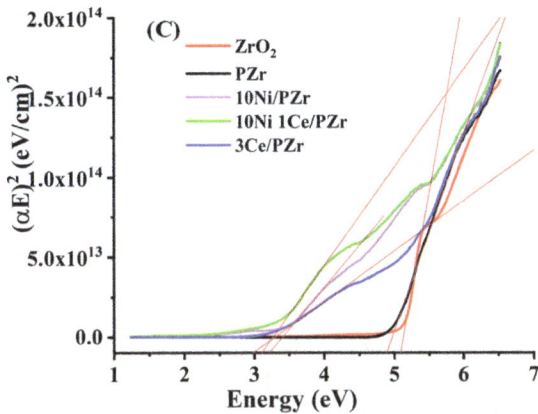

Figure 4.21: The band gap of ZrO_2, PZr, 3Ce/PZr, 10NiPZr and 10Ni1CePZr catalysts. With permission from Wiley.

References

[1]   Yang, L., Kruse, B. *J. Opt. Soc. Am. A* **2004**, *21* (10), 1933.

[2]   Yang, L., Kruse, B., Miklavcic, S. J. *J. Opt. Soc. Am. A* **2004**, *21* (10), 1942.

[3]   A. Escobedo Morales, E. Sanchez Mora, Umapada Pal *Rev. Mex. de Fis.* **2007**, S53, 18–22.

[4]   Kumar, R., Enjamuri, N., Pandey, J. K., Sen, D., Mazumder, S., Bhaumik, A., Chowdhury B. Appl. Catal. A. Gen. 2015, 497, 51–57.

[5]   Al-Fatesh, A. S., Chaudhary, M. L., Fakeeha, A. H., Ibrahim, A. A., Al-Mubaddel, F., Kasim, S. O., Albaqmaa, Y. A., Bagabas, A. A., Patel, R., Kumar, R. *Processes* **2021**, *9* (1), 1–15.

[6]   Tsoncheva, T., Ivanova, L., Rosenholm, J., Linden, M. *Appl. Catal. B Environ.* **2009**, *89* (3–4), 365–374.

[7]   Khatri, J., Fakeeha, A. H., Kasim, S. O., Lanre, M. S., Abasaeed, A. E., Ibrahim, A. A., Kumar, R., Al-Fatesh, A. S. *Int. J. Energy Res.* **2021**, June, 45, 19289–19302.

5 Nuclear magnetic resonance spectroscopy

5.1 Background

Under the influence of strong applied magnetic field, energy of the spin active nucleus (of compound) is split into various spin states. When the frequency of radiofrequency wave of the applied magnetic field matches with the oscillating electric field of precessing spin-active nucleus, radiofrequency was absorbed and transition of electron in various spin state is carried out. The study of absorption of radiofrequency radiation by nuclei is known as nuclear magnetic resonance (NMR). Now, excited nucleus returns to ground state and emits the energy/NMR signal. Further, the NMR signal from native nuclei will interact/couple with spin states of nearby nucleus and finally, it is split into various energy level and received by the detector. The NMR signals (original as well as split signals) give us information about the spin-active nucleus and its closest spin activates neighbors. It becomes a very strong tool for structure elucidation of a compound.

5.2 Instrumentation and working principle

5.2.1 Radiofrequency wave resonance

The sample well placed between applied field is wrapped from inductor coil of tank circuit on the X-direction. In tank circuit (LCR circuit or resistive circuit parallel by capacitor and inductor), resistance generates the potential drop, capacitor stores that amount of energy and inductor (in x direction) builds up the radiofrequency magnetic wave on the expense of energy storage of capacitor in the xz plane. When frequency of radiofrequency wave and oscillating electric field of precessing nucleus matched, the resonance happens and radiofrequency waves are absorbed by the spin-active nucleus. In another word, now nuclear spin interacts with applied magnetic field.[a,b] The adsorbed amount brings change in spin states. Clockwise spin state changes to counterclockwise spinning and counterclockwise spin state changes to clockwise spin and vice versa. After expanding the adsorbed energy, both spin states return to their native states. In the upper example, along with other nucleus, nine excess nuclei in ground state also absorb the energy and change their spin state. When all return to the previous energy spin state, the emitted energy amount from none excess nuclei come out from the sample (called NMR signal) in Y-Z direction.

https://doi.org/10.1515/9783110656480-005

5.2.2 Interaction with the nearby hydrogen spin environment

When NMR signal comes out to the original site's nucleus, it will interact with spin states of nearby nucleus. As if it may couple to one up spin of nearby proton and one down spin, the NMR signal splits into two parts each having same magnitude (1:1). It couples with two up spin, one up one down spin, one down one up spin and two down spin. The NMR signal split into 1 (due to couple with two up spin): 2 (due to couple doubles with one up and one down spin): 1 (due to couple with two down spin). Overall, NMR signal (y-z direction) is split into $n + 1$ peak where n is the number of nearby H. The magnitude of splitting simply follows the pyramid as given below:

$$1:4:6:4:1$$
$$1:3:3:1$$
$$1:2:1$$
$$1:1$$

5.2.3 Detection

Finally, an inductor which wraps the sample in y-axis is able to carry out NMR signal in y-z direction. These signals are converted into relative electric signals, further electric signals are integrated and amplified.

5.2.4 Data interpretation

Silicon is in many respects one of the more important elements in the earth's crust. Among the naturally occurring isotopes ^{28}Si (92.21%), ^{29}Si (4.70%) and ^{30}Si (3.09%), ^{29}Si only has a spin 1/2 and therefore a magnetic moment [1]. So, ^{29}Si NMR spectra have been often used for assigning the chemical environment around the Si atoms in silica-based material.

The majority of ^{29}Si NMR shifts are found in a range between +50 and −200 ppm. It is known that silica has tetrahedral coordination. If four sides of silica are coordinated to all four silica/metal environment, then we get a peak around 110 ppm shift. This peak is abbreviated as Q^4 peak or $(Si/M-O-)_4Si$. Q^4 band may be regarded as continuous sheet. If some defects happen due to calcination or another stress, it can generate defects in continuous sheet. The defective silica environment is shown in Si–NMR as Q^3 and Q^2 band. If three sides of silica are coordinated to another silica/metal environment and one side is coordinated to hydroxyl, then generally, we have got a peak around 100 ppm shift called $Q^3 = (Si/M-O-)_3Si(OH)$ in ^{29}Si-NMR. Again, if two sides of silica are coordinated to two silica/metal environment, a peak of around 90 ppm shift is observed known as $Q^2 = (Si/M-O-)_2Si(OH)_2$ peak (Figure 5.1). So, by combining Q^2, Q^3 and Q^4 peaks' intensity information, chemist can assign the chemical environment around Si atom in silica-based materials. ^{29}Si-NMR spectroscopy allows quantification [2] of silanol groups, by taking all silicon atom of Q^2, Q^3 and Q^4 into consideration, regardless of their location on surface and pore walls.

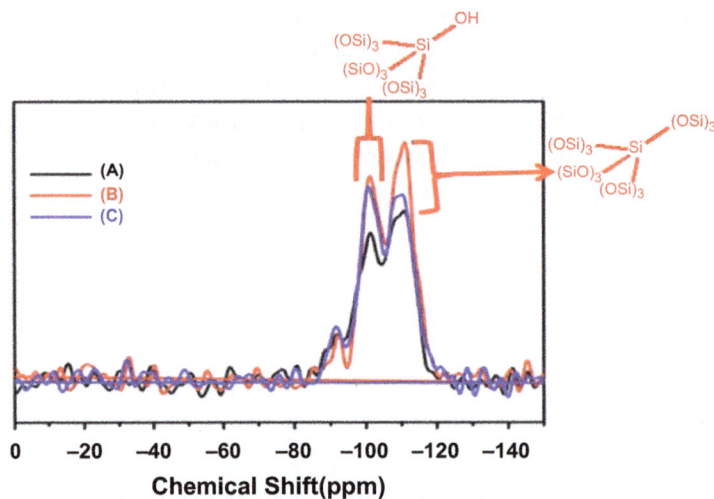

Figure 5.1: A typical ^{29}Si–NMR spectrum [3]. With permission from Elsevier.

Solid-state ^{27}Al MAS NMR put insights about the state and the coordination of the Al species in a material. Signal at 60–50, 40 and 0 ppm corresponding to tetrahedral, pentahedral and octahedral coordination of Al atom, respectively. Apart from ^{27}Al MAS NMR, nowadays, ^{31}P MAS NMR and ^{51}V NMR are used frequently to understand the P and V environment in doped metal oxide chemistry. ^{15}N-NMR is useful technique to understand the linkage of nitrogen functionalized silica/inorganic in synthesized materials. The pure N_2, NH_2–OH, $NH_4{}^+$, Si–NH_2 and NH_3 have ^{15}N-NMR peaks at –72.16, –271, –359, –373 and –381.9 ppm, respectively. Large polarizability, chemical inertness and interaction of useful range of condensed phase (including pure liquids, protein solutions and suspensions of lipid and biological membranes), ^{129}Xe-NMR spectroscopy is used to know the local environment of surface structure and adsorption sites of porous sorbents [4, 5]. Xe in gas phase is shown by 0 ppm chemical shift whereas Xe adsorbed in the pore zeolite system showed higher chemical shift 171–182 ppm.

5.3 Glossary

[a]**Nuclear spin:** Each nucleon (proton or neutron) has 1/2 spin quantum number. Nucleons start to fill in the nuclear orbit in the order of increasing total momentum.

(i) Let us first talk on protons. In the first orbit, first two protons have no angular momentum (say "s" state) and $+\frac{1}{2}$ spin momentum. The total momentum of nucleon is $\frac{1}{2}$ and the energy state is represented $1s_{1/2}$. Every two nucleons are paired up in opposite direction $(1/2 + (-1/2))$ and the total magnetic moment becomes zero.

(ii) In the first orbit, next six protons ($n = 6$) have one angular momentum (say p state) and $+\frac{1}{2}$. The angular momentum and spin momentum of protons will couple each other. More than half protons ($n + 1 = 4$) get total momentum 3/2 $(1 + 1/2)$ as parallel coupling of angular momentum and spin momentum and the energy state is designated as $1p_{3/2}$. Every two nucleons are paired up in opposite direction (3/2 $+ (-3/2)$) and the total magnetic moment becomes zero. The total momentum of rest protons ($6-4 = 2$) is $\frac{1}{2}$ as antiparallel coupling of both momentum and the energy state is designated as $1p_{1/2}$. Every two nucleons are paired up in opposite direction $(1/2 + (-1/2))$ and again the total magnetic moment becomes zero.

(iii) In this way, protons as well as neutrons may be filled in the given order. Simply pick up the energy state terms from head to tail direction and fill up the protons. In the same way, fill up the neutrons separately. After filling all protons of atom, if protons remain unpaired, it contributes total momentum to the atom. Some unpaired neutron contribute toward total momentum. The total momentum of nucleus (of atom) or spin magnetic moment (I) is the sum of the momentum due to protons and neutrons.

[b]Interaction of nuclear spin with applied magnetic field (by bar magnet):

When a sample which has to be analyzed is placed between the magnetic bar or applied magnetic field, magnetic spin of nucleus interacts with the applied magnetic field in the following ways.

(i) Under the influence of magnetic field, spinning nucleus (around its axis) start to align about applied magnetic field (Figure 5.2). As per the alignment, spin of nucleus quantizes into different states ranging from I, $(I-1)$, . . ., till $-I$. The separation of energy state is directly proportional to applied magnetic field and magnetogyric ratio

$$E = \gamma Bh/2\pi$$

Apart from alignment and quantization of state, every nucleus starts to wobble or precess around the applied magnetic field with angular frequency ω.

Overall, 1.41 T magnetic field can generate 60 MHz precessional frequency (ω) of "H" nucleus. 60 MHz precessing nucleus can generate oscillating electric field having frequency 60 MHz. For collection of atoms, all the spin states are populated and each generates oscillating electric field of 60 MHz. The number of nuclei in higher energy states can be calculated by Boltz distribution law as

$$N/No = e^{-E/kt}$$

If $N/N_o = 100,000/100,009$, it indicates that ground level has nine excess nuclei than excited state.

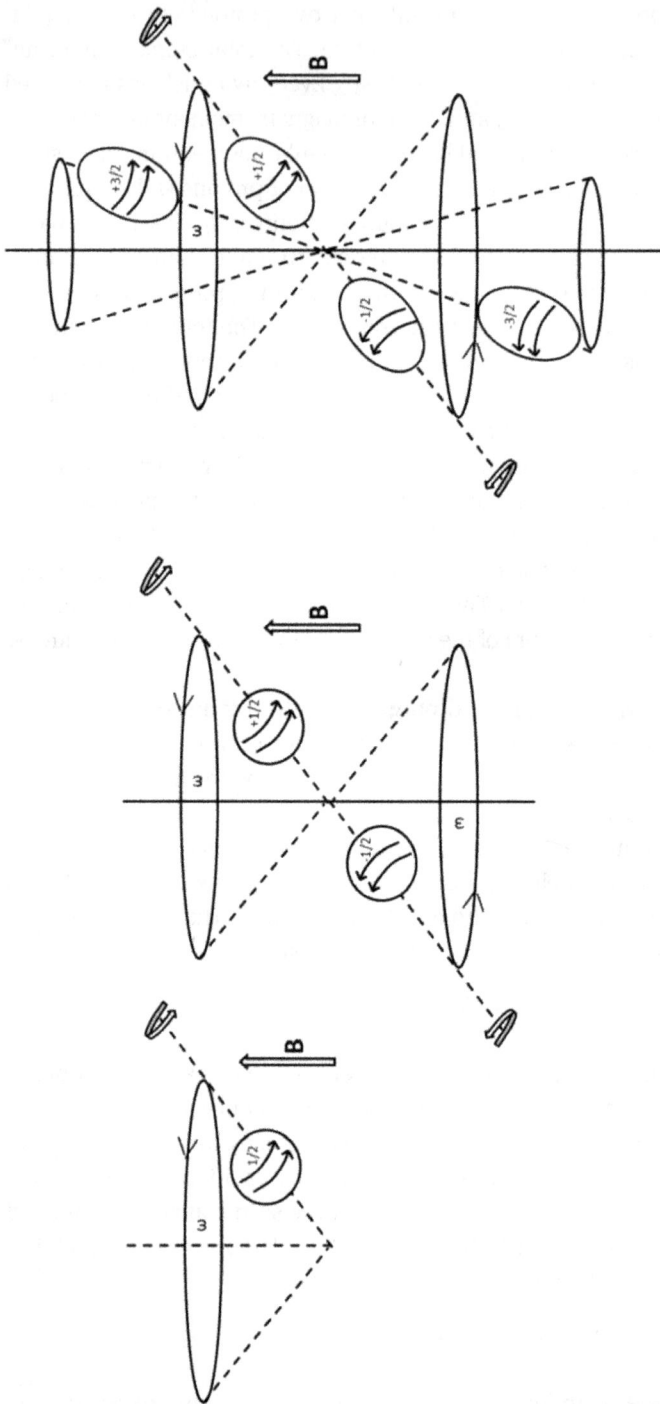

Figure 5.2: Alignment of spinning nucleus (around its axis) under the influence of magnetic field.

5.4 Analysis of ^{11}B, ^{27}Al, ^{29}Si, ^{31}P, ^{51}V and NMR profile

Example 1: Zhang et al. [6] prepared phosphorous-impregnated (heating salt precursor over a support till dryness) aluminosilicates using phosphoric acid solution as phosphorous precursor [6]. The silicates have ordered mesoporous material with sustainable acidity and they are abbreviated as Al–MCM-41 (Si/Al = 20) and 3% P-impregnated Al–MCM-41 (Si/Al = 20). The ^{27}Al and ^{31}P NMR spectra of Al–MCM-41 (Si/Al = 20) and 3% P-impregnated Al–MCM-41 (Si/Al = 20) are shown in Figure 5.3. Only aluminosilicates has Al–NMR peak at 53 and 0 ppm, respectively, for tetrahedral (framework) and octrahedral (extra framework) aluminum respectively. At phosphorous, impregnated aluminosilicates have additional peak at 42 and −14 ppm indicates aluminum interaction with some species during phosphorous introduction (left; Figure 5.3a). Yuan et al. [7] assigned the signal around 42 ppm for tetrahedral aluminum bonded to phosphorus atoms via oxygen bridges (P–O–Al) whereas −14 ppm peak is attributed to octahedral aluminum bonded to phosphorus (left; Figure 5.3b). ^{31}P MAS NMR spectra showed more clear idea about phosphorous interaction with other species. A broad peak at −23 ppm in ^{31}P MAS NMR is attributed to P–O–Si or Si–O–P–O–Al structures (right; Figure 5.3).

Example 2: Anilkumar and Hölderich [8] have prepared Nb-incorporated aluminosilicates (MCM-41) by using sodium metasilicates as a source of silica and niobium petachloride as a source of niobia and tetradecyltrimethyl ammonium bromide (TDTAB) as template [8]. The molar composition of synthesis get at pH 10.5 were as follow SiO_2:$NbCl_5$:TDTAB:H_2SO_4:H_2O: x: y: 0.2: 0.89: 120 (x/y = Si/Nb = 128–16). The finally dried and calcined powder was named as Nb-MCM-41. The ^{29}Si NMR spectra of as-synthesized MCM-41, as-synthesized Nb–MCM-41 (Si/Nb = 32) and calcined Nb–MCM-41 (Si/Nb = 32) are shown in Figure 5.4. It was found that with increasing niobium loading (Si/Nb = 128 to 16) in Nb–MCM-41, Q^3 (($Si-O-)_3Si-OH$) density increases on expanse of decrease in Q^2 (($Si-O-)_2Si-(OH)_2$) in ^{29}Si–NMR. It indicates the growth of acid site with rise of silanol sites with increasing metal loading.

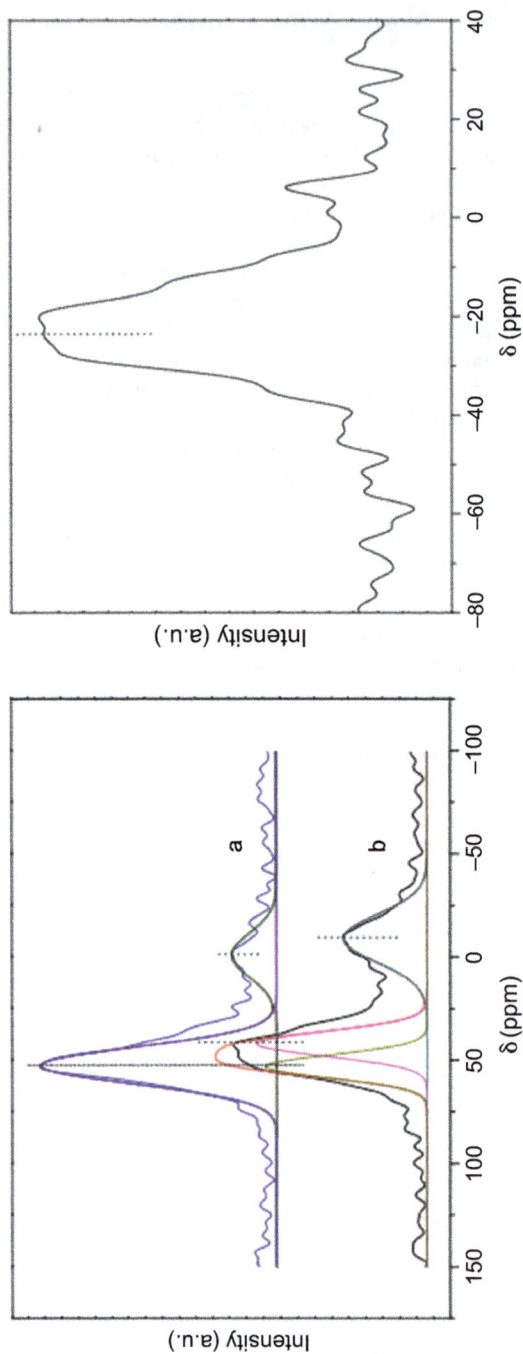

Figure 5.3: Left (a): The ^{27}Al spectra of Al–MCM-41 (Si/Al = 20) and (b) 3% P-impregnated Al–MCM-41 (Si/Al = 20). Left: ^{31}P–NMR spectra of 3% P-impregnated Al–MCM-41 (Si/Al = 20). With permission from Elsevier.

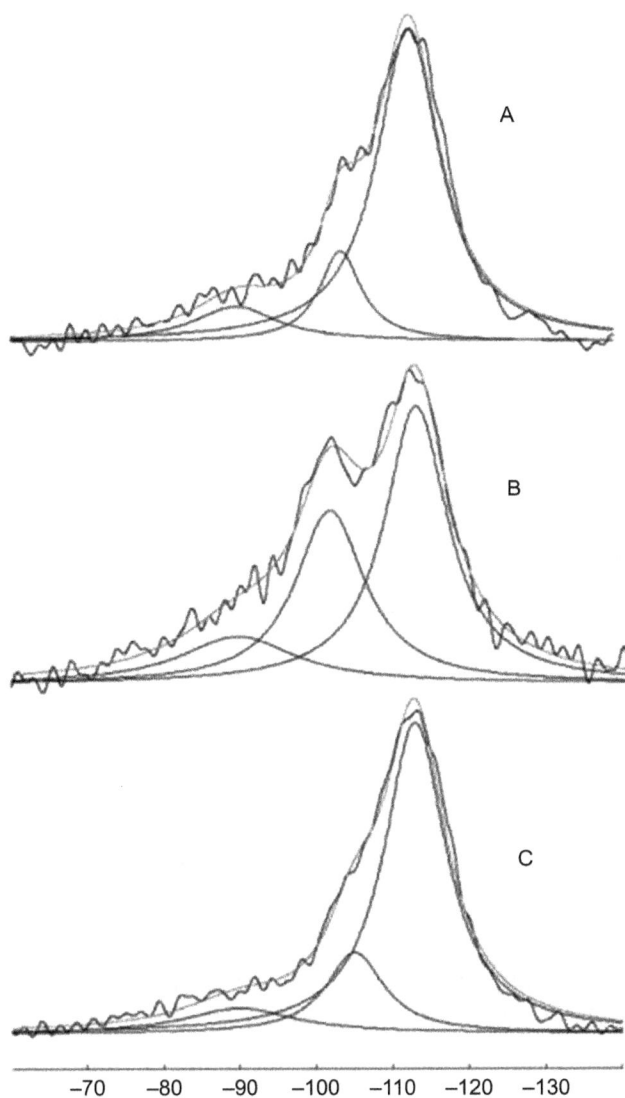

Figure 5.4: The ^{29}Si NMR spectra of (A) as-synthesized MCM-41, (B) as-synthesized Nb-MCM-41 (Si/Nb = 32) and (C) calcined Nb-MCM-41 (Si/Nb = 32). With permission from Elsevier.

Example 3: Forni et al. [9] have prepared boria–alumina high surface area catalyst through sol–gel procedure using butanol solution of $(NH_4)_2B_4O_7 \cdot 4H_2O$ and butane-1,3-diol solution of aluminum-tri-sec-butoxide and water in proportional amount [10, 9]. The material is B_2O_3–Al_2O_3 (B/Al = 0, 0.013, 0.062, 0.184, 0.728). The ^{27}Al and ^{11}B NMR spectra of B_2O_3–Al_2O_3 (B/Al = 0, 0.013, 0.062, 0.184, 0.728) catalyst sample are shown in Figures 5.5 and 5.6 respectively. After proper hydrolysis, mixture was evaporated, dried and calcined. ^{27}Al–MAS–NMR analysis showed peaks at 60, 40 and 0 ppm which indicated the presence of tetrahedral-, pentahedral- and octahedral-coordinated aluminum in the sample respectively. As boron loading increases, pentahedral-coordinated aluminum amount increases. ^{11}B–MAS–NMR showed peaks at 17.5 and 1 ppm for trigonal BO_3 and tetrahedral BO_4 species respectively. Again, with increasing boron loading, tetrahedral BO_4 species became more prominent. It was expected that trigonal BO_3 were incorporated in the lattice whereas tetrahedral BO_4 formed during rehydration of the calcined samples at catalyst surface.

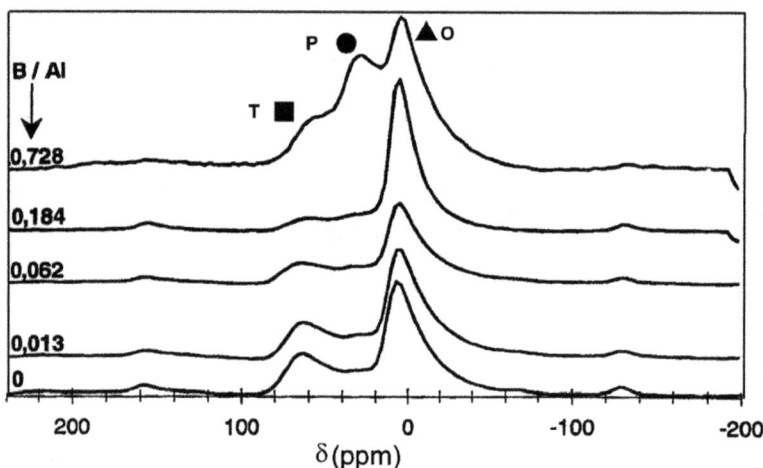

Figure 5.5: The ^{27}Al spectra of B_2O_3–Al_2O_3 (B/Al = 0, 0.013, 0.062, 0.184, 0.728) catalyst. Dark square, triangle and circle are for tetra-coordinated aluminum, octa-coordinated aluminum and penta-coordinated aluminum, respectively. With permission from Elsevier.

Example 4: Pillai et al. had prepared vanadium-doped tin oxide by impregnating oxalic acid solution of ammonium metavanadate over SnO_2. The SnO_2 itself was prepared by adding ammonia solution and stannic chloride solution at pH 4–5, followed by filtration, washing, drying and calcination [11]. The final catalyst material was termed as V_2O_5/SnO_2. ^{51}V–NMR spectra of V_2O_5/SnO_2 material is shown in Figure 5.7. ^{51}V–NMR of 6–12 wt% V_2O_5/SnO_2 samples showed peak at 150.3 and 257.9 ppm for tetrahedral vanadium species and octahedral vanadium species. As vanadia loading increased octahedral vanadia amount was dominated (upper NMR spectra of Figure 5.3). It indicated that at low vanadia loading, vanadia was well dispersed as monolayer tetrahedral units. When vanadia loading exceeded more than monolayer, it was well crystallized on the catalyst surface in octahedral vanadia unit.

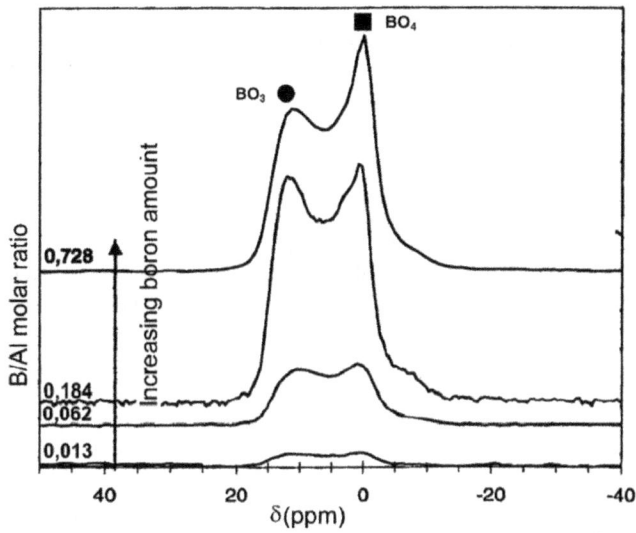

Figure 5.6: ^{11}B NMR spectra of B_2O_3–Al_2O_3 (B/Al = 0, 0.013, 0.062, 0.184, 0.728) catalyst. Dark square and circle are for BO_4 species and BO_3 species respectively. With permission from Elsevier.

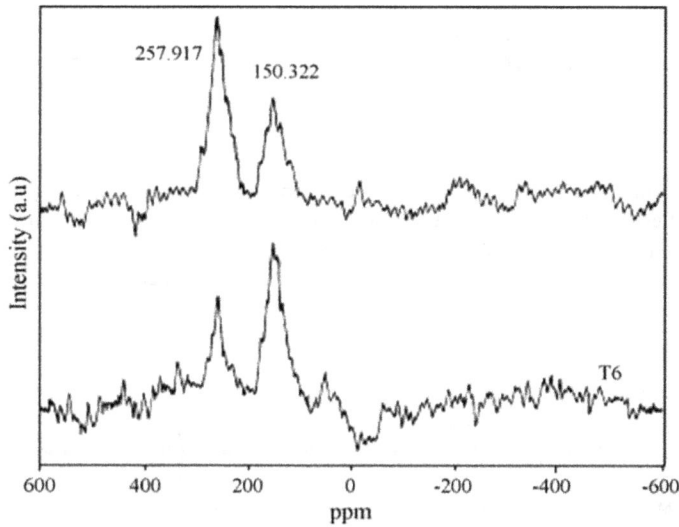

Figure 5.7: ^{51}V NMR spectra of V_2O_5/SnO_2 material. With permission from Elsevier.

Example 5: Wang et al. [12] had prepared hexadecyl trimethyl ammonium phosphomolybdate by mixing of the ethanolic solution of tetramethyl ammonium chloride and phosphomolybdic acid and followed by vacuum drying at 80 °C for 10 h [12, 13]. $[n\text{-}C_{16}H_{33}N(CH_3)_3]H_2PW_{12}O_{40}$ catalyst was abbreviated as CTAH$_2$PW. The catalyst was employed for oxidation of cyclohexanol to cyclohexanone reaction. The ^{31}P–NMR spectra of fresh catalyst, catalyst recovered after oxidation state and catalyst recovered after Beckmann rearrangement reaction are shown in Figure 5.8. ^{31}P–NMR of fresh catalyst showed peak at –15.6 ppm for $[PW_{12}O_{40}]^{3-}$ whereas catalyst recovered in oxidation stage showed two peaks, one at –15.6 ppm for $[PW_{12}O_{40}]^{3-}$ and another at –1.2 ppm for active tungsten–peroxo complex. The peak for tungsten–peroxo complex was again not found for catalyst recovered after the Beckmann rearrangement reaction. It indicates that tungsten–peroxo complex is the main catalytic-active species which was formed and used up during the reaction.

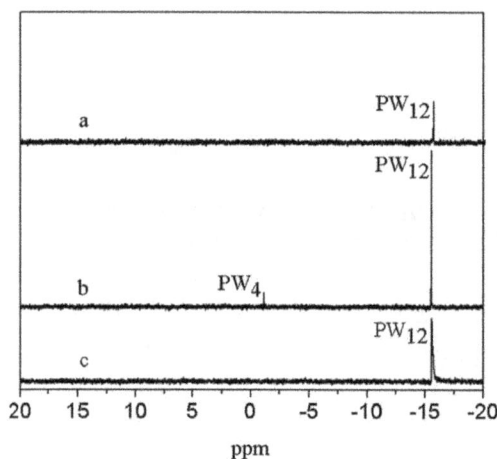

Figure 5.8: ^{31}P NMR spectra of (a) fresh catalyst, (b) catalyst recovered after oxidation state and (c) catalyst recovered after the Beckmann rearrangement reaction. PW$_{12}$ and PW$_4$ are abbreviated for $[PW_{12}O_{40}]^{3-}$ and for tungsten–peroxo complex. With permission from Elsevier.

References

[1]   C. Brevard, P. Granger. *"Handbook of High Resolution Multinuclear NMR"*, **1982**. Wiley: New York ISBN: 9780471063230

[2]   Palkovits, R., Yang, C. M., Olejnik, S., Schüth, F. *J. Catal.* **2006**, *243* (1), 93–98.

[3]   Kumar, R., Shah, S., Bahadur, J., Melnichenko, Y. B., Sen, D., Mazumder, S., Vinod, C. P., Chowdhury, B. *Microporous Mesoporous Mater.* **2016**, *234*, 293–302.

[4]   Nossov, A., Haddad, E., Guenneau, F., Mignon, C., Gédéon, A., Grosso, D., Babonneau, F., Bonhomme, C., Sanchez, C. *Chem. Commun.* **2002**, *21*, 2476–2477.

[5]   Bekyarova, E., Kaneko, K., Yudasaka, M., Kasuya, D., Iijima, S., Huidobro, A., Rodriguez-Reinoso, F. *J. Phys. Chem. B* **2003**, *107* (19), 4479–4484.

[6]   Zhang, D., Wang, R., Yang, X. *Catal. Commun.* **2011**, *12* (6), 399–402.

[7]   Yuan, Z. Y., Chen, T. H., Wang, J. Z., Li, H. X. *Colloids Surfaces A Physicochem. Eng. Asp.* **2001,** *179* (1–3), 253–259.

[8]   Anilkumar, M., Hölderich, W. F. *J. Catal.* **2008,** *260* (1), 17–29.

[9]   Dumeignil, F., Guelton, M., Rigole, M., Amoureux, J. P., Fernandez, C., Grimblot, J. *Colloids Surfaces A Physicochem. Eng. Asp.* **1999,** *158* (1–2), 75–81.

[10] Forni, L., Fornasari, G., Tosi, C., Trifirò, F., Vaccari, A., Dumeignil, F., Grimblot, J. *Appl. Catal. A Gen.* **2003,** *248* (1–2), 47–57.

[11] Salinas, D., Escalona, N., Pecchi, G., Fierro, J. L. G. *Fuel* **2019,** *253* (January), 400–408.

[12] Wang, H., Hu, R., Yang, Y., Gao, M., Wang, Y. *Catal. Commun.* **2015,** *70*, 6–11.

[13] Qiu, J., Wang, G., Zeng, D., Tang, Y., Wang, M., Li, Y. *Fuel Process. Technol.* **2009,** *90* (12), 1538–1542.

6 X-ray diffraction (XRD)

6.1 Background

When X-ray is glancing over an atom, its energy is absorbed by the atom. Very soon, electrons of atom scatter the energy. In a crystal, atoms are separated by regularly specified distances and make ordered atomic plane. If the ordered atomic planes are exposed to X-ray beam at incident angle θ, the X-ray energy is absorbed and very soon emitted to the other side at very specific angle (again θ). The angle between incident angle and scattering angle is 2θ. Different beams reflecting from same plane at different depth reach to the detector. It has different phase/path difference as per the different depths of same plane. If path difference of rays are in the order of $n\lambda$ (where λ is wavelength of X-ray), these will interfere constructively and wave is amplified and detected by detector. Again, if path difference of rays is in the order of $(n/2)\lambda$, these waves are completely out of alignment and so interfere destructively, or simply wave/signal is destroyed and not detected by detector. In the same way, detector records signal coming from different plane. The data recorded in the form of intensity vis-à-vis 2θ.

6.2 Instrumentation and working principle

X-ray is a powerful technique to examine the periodicity/crystallinity of the sample. The schematic diagram of X-ray diffractometer is shown in Figure 6.1. First, sample is taken in powder form in rotating sample holder so that all planes of sample will be exposed to X-ray beam. Now, sample is irradiated with $Cu_{k\alpha}$ X-rays. Electron beam is emitted from cathode (through either thermionic or field emission), accelerated by applied voltage (between cathode and anode) and stabilized against voltage fluctuation by Wehnelt electrode. High-energy electrons strike Cu metal anode and generate $Cu_{k\alpha}$ X-rays (energy = 8.04 keV). Electron gun along with striking anode is called X-ray gun. Samples irradiated with $Cu_{k\alpha}$ X-rays are elastically scattered by atom is periodic lattice in different successive layer. In that mean path difference (phase difference) is created in reflected beam.

The scattered convergent beam is collected by detector (placed in diffracted beam path). The scattered monochromatic X-rays that are in phase give constructive interference and are detected by detector. By Bragg's law[a] ($n\lambda = 2d\sin\theta$), we are able to drive the lattice spacing. The detector gives intensity versus $2\theta^*$ data and the corresponding graph. With the help of crystal library, reciprocal planes for each interplanar spacing[b] (hkl) is calculated. Using $1/d^2 = h^2/a^2 + k^2/b^2 + l^2/c^2$ formula (law of rational indices[c]) for different sets of plane and corresponding interplanar spacing (d-spacing), lattice parameter (a, b, c) can be calculated. By analyzing lattice parameter with angles, we come to know the information phase present in the crystal (crystal system[d]).

https://doi.org/10.1515/9783110656480-006

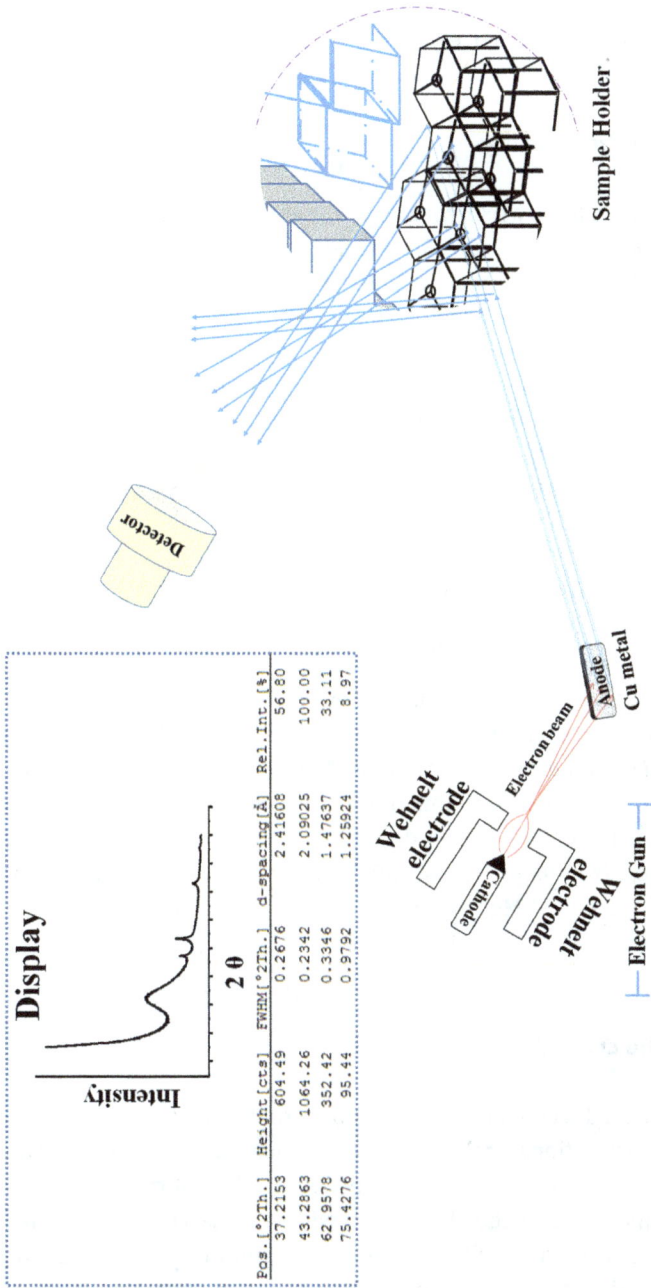

Figure 6.1: The schematic diagram of X-ray diffractometer.

6.3 Data correlation

The peak shift variance in (d-spacing) and peak broadening in X-ray diffraction (XRD) profile are correlated with fitting of different sized atom in lattice and in homogeneous strain within a crystallite respectively.

6.3.1 Peak shift variance in "*d*" spacing

By comparing sample data with library reference data/closely related samples, we find deviation in interplanar spacing "*d*" (Figure 6.3). A side-by-side comparison of XRD data from sample and data from crystal library of cubic NiO are shown in Figure 6.2.

The variance in *d* spacing can be argued in two ways. In the case of stretching, planes are distant to each other. Spacing expands from *d* to (*d* + *δd*). Then, by Bragg's law, the position of the peak will decrease from *θ* to (*θ*–*δθ*). In the case of compression, plane comes closer to each other, and so interplanar spacing contracts from *d* to (*d*–*δd*). Then, by Bragg's law, the position of the peak will increase from *θ* to (*θ* + *δθ*). Then, the original and modified Bragg's equation can be shown as

$$n\lambda = 2d\sin\theta$$

$$n\lambda = 2(d + \delta d)\sin(\theta - \delta\theta)$$

$$n\lambda = 2(d - \delta d)\sin(\theta + \delta\theta)$$

This shift may be due to fitting of different sized atom in lattice. It may be due to doping of larger cation in place smaller cation (radius of Ce^{4+} = 1.098 Å). As Santra et al. [1] found that the *d* spacing of CeO_2 is 3.1106 Å, doping with 4 mol % Ba^{2+} (radius of Ba^{2+} = 1.49 Å) in CeO_2 lattice (radius of Ce^{4+} = 1.098 Å) causes expansion of "*d*" spacing to 3.1153 Å. It may be due to doping of smaller cation in place of comparatively larger cation (radius of Ce^{4+} = 1.098 Å). Santra et al. [1] found that doping with 4 mol% Mg^{2+} (radius of Mg^{2+} = 0.86 Å) in CeO_2 lattice (radius of Ce^{4+} = 1.098 Å) causes contraction of "*d*" spacing to 3.1153 Å.

6.3.2 Peak broadening and crystalline size

Homogeneous atomic layer produces sharp diffraction pattern. Structural defects (such as interstitials, vacancies, dislocations and layer faults) induce inhomogeneous strain within a crystallite. If system is strained nonhomogeneously, different crystallites will be strained by different amounts and the shifts in 2*θ* will be variable. This is called peak broadening and it is measured as full width at half magnitude (*β*). A systematic representation of shifts in 2*θ* and peak broadening due to inhomogeneous strain is shown in Figure 6.3.

X-ray diffraction Data from Sample

Pos. [°2Th.]	Height [cts]	FWHM [°2Th.]	d-spacing [Å]	Rel. Int. [%]
37.2153	604.49	0.2676	2.41608	56.80
43.2863	1064.26	0.2342	2.09025	100.00
62.9578	352.42	0.3346	1.47637	33.11
75.4276	95.44	0.9792	1.25924	8.97

Data from crystal Library of Cubic NiO (Ref: 00-001-1239)

h	k	l	d [Å]	2Theta [deg]	I [%]
1	1	1	2.40000	37.442	60.0
2	0	0	2.08000	43.473	100.0
2	2	0	1.47000	63.204	60.0
3	1	1	1.26000	75.374	24.0
2	2	2	1.20000	79.870	12.0

Figure 6.2: A side-by-side comparison of X-ray diffraction data from sample and data from crystal library of cubic NiO.

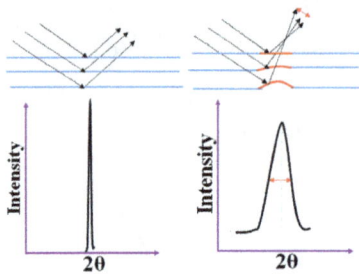

Figure 6.3: A systematic representation of peak broadening due to inhomogeneous strain in the sample.

Apart from vacancy, doping of lower valent cation (either large[2] or small radius[3]) causes decrease of full width at half magnitude (β) and so decrease of particle size later. The crystalline size (t) is calculated by the Scherrer equation[e]($t = K\lambda / \beta \cos\theta$)

About 10 mol% Sm doping in CeO_2 lattice results in smaller crystalline size than pure CeO_2. It may be due to the decrease in concentration of surface hydroxyls as low valent samarium (but larger radius Sm^{3+} = 0.1219 nm) replaces the higher valent ceria in the lattice [1] (Figure 6.4; left). In another example, 4 mol% Co doping in CeO_2 lattice gives smaller crystallite due to the decrease in concentration of surface hydroxyls as low valent cobalt (smaller radius Co^{2+} = 0.075 nm, Co^{3+} = 0.061 nm) replaces the higher valent ceria in the lattice [2] (Figure 6.4; right).

6.4 Glossary

[a]**Bragg's law:** Let us suppose a ray is glancing parallel to the plane and there is an imaginary perpendicular line to this ray (Figure 6.5A). Now, rays glance with angle θ to the horizontal plane, the imaginary perpendicular line also makes θ to the depth of the plane or interplanar spacing "d." Some rays are reflected from top plane and some of them penetrate down and reflect from successive layers (Figure 6.5B, C). Now, reflected rays from inner layer have path difference than upper plane reflected rays. So, upper layer reflected rays march forward than lower plane reflected rays having path difference ΔX.

The path difference ΔX_i can be calculated geometrically, the path difference in incident rays is $\Delta X_i = d \sin\theta$. The path difference in reflected rays is $\Delta X_r = d \sin\theta$. The total path difference is $\Delta X = 2d \sin\theta$. Angle between reflected rays and transmitted rays will 2θ. The detector has to receive the rays of these 2θ regions. Detector receives beams having path differences. It may be the following two types. If the path difference between two rays are in the order of X-ray wavelength λ as λ, 2λ, 3λ, . . ., $n\lambda$ ((Figure 6.5D), then it creates constructive pattern and it is detected by detector. Then, path difference $\Delta X = n\lambda$; $n = 1, 2, ...$ Now, total path difference can be shown as $n\lambda = 2d \sin\theta$. It is also known as Bragg's equation. By this interplanar spacing, "d" can be found.

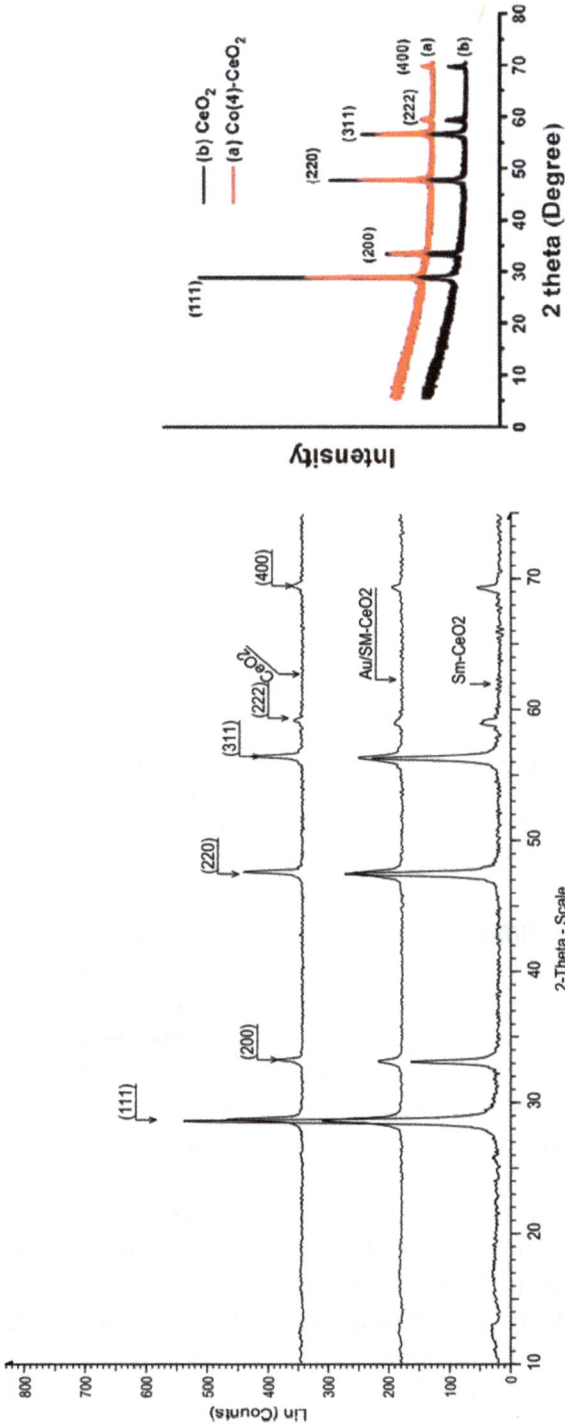

Figure 6.4: Left: XRD pattern of CeO_2, Sm-doped CeO_2 (Sm/Ce = 4/100) and Au-deposited Sm-doped CeO_2. **Right:** X-ray diffraction patterns of (a) CeO_2 (b) Co(4)–CeO_2 (Co is 4 mol%). With permission from Elsevier.

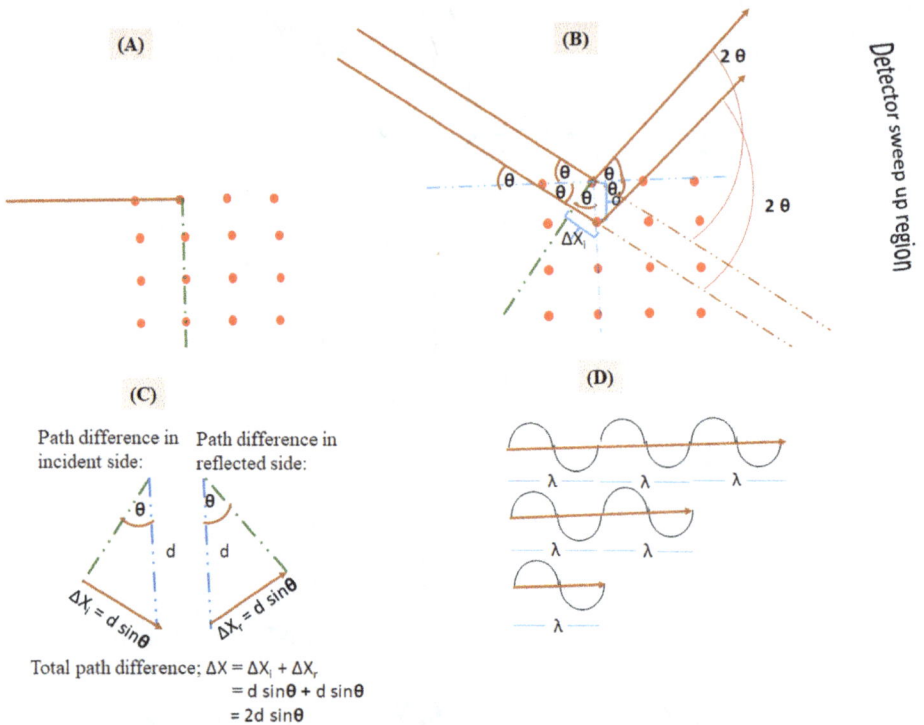

Figure 6.5: (A) Glancing of ray parallel to the plane. (B) Glancing of ray at angle θ to the horizontal plane. A path difference "ΔX" is created between reflected beam on successive plane.
(C) Interplanar distance "d" is related to path difference "ΔX" and glancing angle "θ." (D) The path difference "ΔX" between reflected rays (from successive plane) is equal to $n\lambda$.

[b]**Interplanar spacing (hkl):** After diffraction from periodic lattices, both transmitted and diffracted X-ray beams fall on screen (Figure 6.6A). If X-ray is diffracted from origin of lattice (000), then it lands on infinite ($\infty\infty\infty$) or it can be said that if does not land on screen, it leaves the screen parallel. Now, in a different way, we can say that, if X-ray lands on the screen parallel ($\infty\infty\infty$) or if it does not land on screen, then that means light is diffracted from atom points at origin (000). The coordinate of light on screen is called reciprocal space and the coordinate of atom is called space.

That means, if X-ray lands on screen at reciprocal space or Miller indices (222), then light is diffracted from atoms points at space ($\frac{1}{2}$, $\frac{1}{2}$, $\frac{1}{2}$) (Figure 6.6B). One can find the information of different reciprocal space (or Miller indices, hkl) against interplanar spacing, 2θ and peak intensity percentage through matching with library database (Figure 6.6C).

(B)

(222)

d

Diffracted ray

θ

$(\gamma_2, \gamma_2, \theta_2)$

Transmitted ray

(000)

(C)

Ref. Pattern:Bursente,00-00-1239

Intensity [%]

100

50

0

40 50 60 70 80 90 100 110 120 130 140

Position [°2 Theta]

h	k	l	d [A]	2Theta [deg]	I [%]
1	1	1	2.40000	37.442	60.0
2	0	0	2.08000	43.473	100.0
2	2	0	1.47000	63.204	60.0
3	1	1	1.26000	75.374	24.0
2	2	2	1.20000	79.870	12.0

(A)

(h,k,l)

d_{hkl}

Diffracted ray

θ

θ

θ

(000)

Transmitted ray

Incident ray

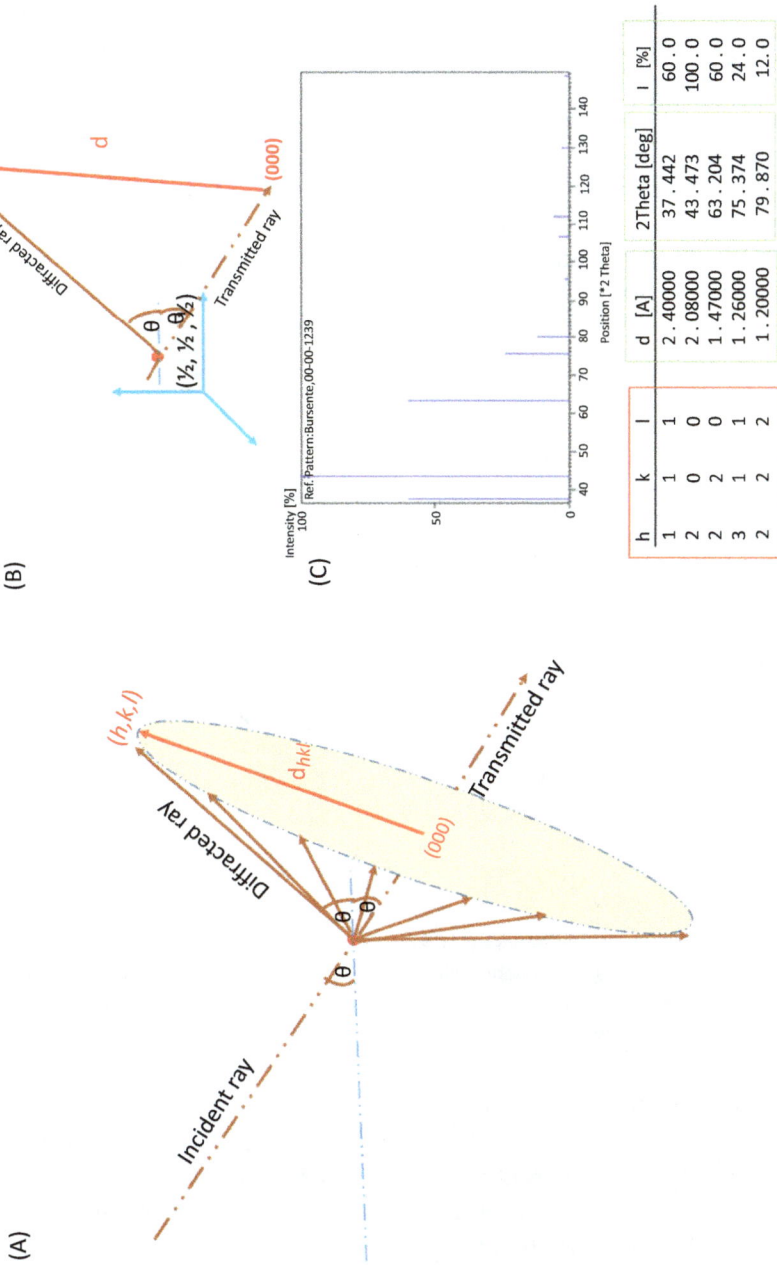

Figure 6.6: (A) Fall of diffracted rays on screen: Space (coordinated of atom) and reciprocal space (coordinated of light on screen). (B) Example of reciprocal space and space. (C) A typical library database and diffraction line pattern of a crystal.

cLaw of rational indices: The mathematical reciprocal of Miller indices (*hkl*) simply gives the atomic points of diffraction. It is treated as unit intercept on crystallographic axis as 1/*h* on X-axis, 1/*k* on Y-axis and 1/*l* on Z-axis (Figure 6.7A). The intercept of any face to crystal along the crystallographic axis is product of side of unit cell and unit intercept. As "a" x "1/*h*" = a/*h* on X-axis, "b" x "1/*k*" = b/*k* on Y-axis and "c" x "1/*l*" = c/*l* on Z-axis (Figure 6.7B, C). Now, it is assumed that same type of plane also passes from origin. So, interplanar spacing (d) can be shown as a vertical line between two planes or a line passing through origin at one end (O) and touching another plane ABC at T from another end. OT is perpendicular to plane ABC.

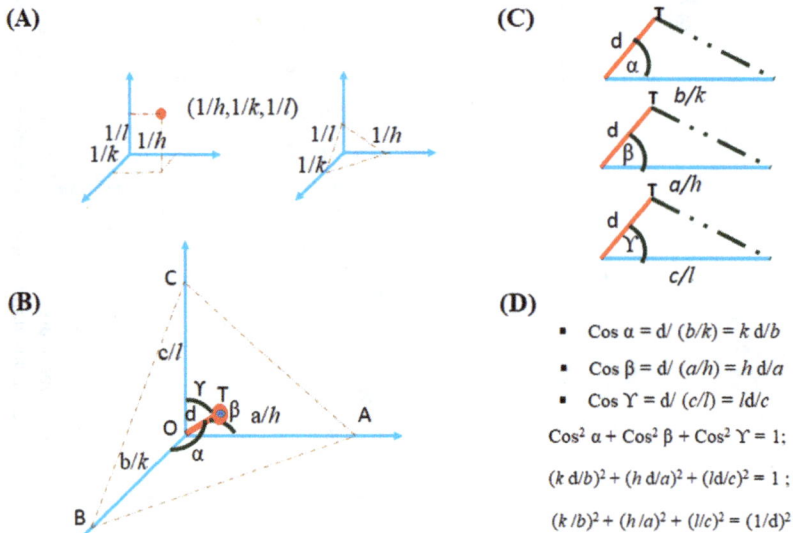

Figure 6.7: (A) Reciprocal of Miller indices (*hkl*) as unit intercept on crystallographic axis. (B) The intercept of any face to crystal equivalence with product of side of unit cell and unit intercept. (C) Every intercept has certain leaning angle (as *α*) with interplanar spacing "*d*." (D) The relation of leaning angle with intercept and interplanar spacing.

Let us suppose that interplanar spacing "*d*" makes angle "*α*" with intercept b/*k*, then, Cos*α* = base/hypotenuse = d/(b/k) = k d/b (Figure 6.7D). Same interplanar spacing "*d*" makes angle "*β*" with intercept a/*h*, then Cos*β* = base/hypotenuse = d/(a/h) = h d/a. In the same way, interplanar spacing "*d*" make angle "Y" with intercept a/*h*, then cosY = base/hypotenuse = d/(c/l) = ld/c.

By trigonometry, we know that $Cos^2\alpha + Cos^2\beta + Cos^2\Upsilon = 1$. After putting the value of Cos*α*, Cos*β* and CosY, relation between *d* spacings and Miller indices can be found as

$$(k/b)^2 + (h/a)^2 + (l/c)^2 = (1/d)^2 \tag{6.1}$$

[d]**Crystal system:** On a particular 2θ, take (hkl) value from library database (Figure 6.8) and take "d" value from detector. After putting the data of hkl and d spacing in relation (1), three equations can be derived. These three equations are enough to solve the three-lattice parameter a, b and c. By knowing lattice parameters as side and angles, crystal phase can be concluded.

From library			From sample XRD file				
h	k	l	Pos. [°2Th.]	Height [cts]	FWHM [°2Th.]	d-spacing [Å]	Rel.Int.[%]
1	1	1	37.2153	604.49	0.2676	2.41608	56.80
2	0	0	43.2863	1064.26	0.2342	2.09025	100.00
2	2	0	62.9578	352.42	0.3346	1.47637	33.11
3	1	1	75.4276	95.44	0.9792	1.25924	8.97

Figure 6.8: A typical database from library database.

$$(1/b)^2 + (1/a)^2 + (1/c)^2 = (1/2.416)^2$$

$$(2/b)^2 + (0/a)^2 + (0/c)^2 = (1/2.090)^2$$

$$(2/b)^2 + (2/a)^2 + (0/c)^2 = (1/1.476)^2$$

Based on types of side and angle between side seven crystals are defined name cubic, tetragonal, orthorhombic, hexagonal, rhombohedral, monoclinic and triclinic (Figure 6.9)

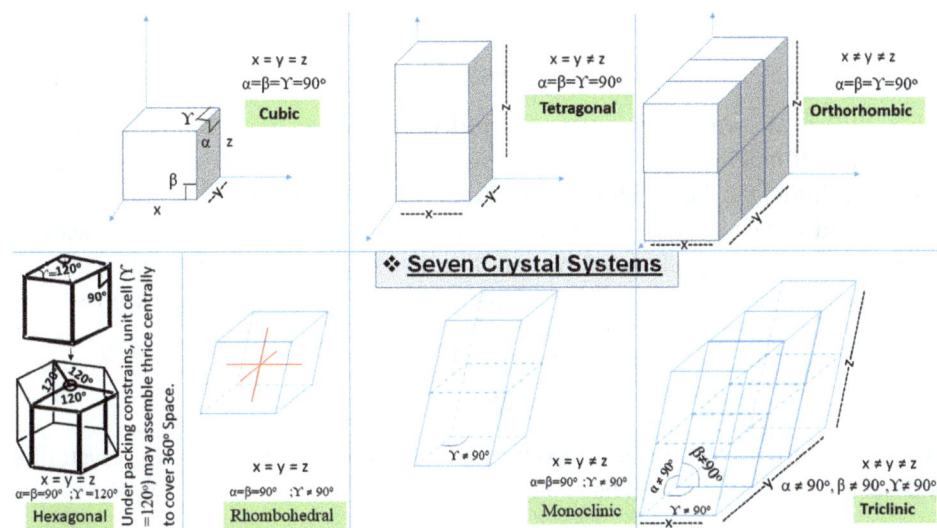

Figure 6.9: The seven crystal systems.

eScherrer's equation: In the Bragg's equation, in place of peak shift $\delta\theta$, full width at half magnitude (β) can be incorporated. This is Bragg's equation of broadening $n\lambda = 2(d - \delta d)\sin(\theta + \beta)$. Strain ($\epsilon$) is defined as ratio of relative change of "d" spacing. Strain (ϵ) = $\delta d/d$. "δd" can be written as product of strain (ϵ) and "d" spacing. So, Bragg's equation of broadening can be written as

$$n\lambda = 2(d - \epsilon d)\sin(\theta + \beta)$$

$$n\lambda = 2(d - \epsilon d)(\sin\theta \cos\beta + \cos\theta \sin\beta) \tag{6.2}$$

Full width at half magnitude (β) cannot be more than 1. So, $\cos\beta \sim 1$ and $\sin\beta \sim \beta$:

$$n\lambda = 2(d - \epsilon d)(\sin\theta + \beta\cos\theta)$$

$$n\lambda = 2d\sin\theta - 2\epsilon d\sin\theta + 2d\beta\cos\theta - 2\epsilon d\beta\cos\theta \tag{6.3}$$

After putting $n\lambda$ in place of $2d\sin\theta$ (Bragg's equation, $n\lambda = 2d\sin\theta$), the equation becomes

$$n\lambda = n\lambda - 2\epsilon\, d\sin\theta + 2d\beta\cos\theta - 2\epsilon d\beta\cos\theta$$

$$\epsilon\sin\theta + \epsilon\beta\cos\theta = \beta\cos\theta \tag{6.4}$$

Dividing all by $\cos\theta$

$$\epsilon\tan\theta + \epsilon\beta = \beta$$

$$\beta = \epsilon\tan\theta/(1 - \epsilon) \sim 4\epsilon\tan\theta$$

Broadening (β) is shown by changing 2θ or $\Delta(2\theta)$ or angular width:

$$\beta = \Delta(2\theta) = 4\epsilon\tan\theta \tag{6.5}$$

With XRD peak, we can calculate the full width at half magnitude/half height. Reach at half height of XRD peak. The width of peak at that height is full width at half magnitude. Further, with eq. (6.5), strain (ϵ) can be calculated.

Now, if peak becomes broad, the sharpness/intensity of the peak decreases. It causes decrease in crystallite size. We can calculate the crystalline size by taking derivative of Bragg's equation with respect to interplanar distance "d" and Bragg's angle "θ"

$$\lambda = 2d\sin\theta$$

$$\lambda = 2\Delta d\cos\theta\Delta\theta$$

After putting peak width half maxima (β) at $2\Delta\theta$ and thickness "t" (or crystallite size) at Δd, the equation will be

$$\lambda = \Delta\, d\beta\cos\theta$$

$$\lambda = t\beta\cos\theta$$

$$t = \lambda/\beta\cos\theta$$

The upper equation is for maximum symmetry (called shape factor 1) as ideal circle, sphere, square or cube. But practically, it may not possible. So, a shape factor having typical value 0.9 is used in upper equation.

$$t = K\lambda/\beta\cos\theta$$

6.5 Analysis of XRD profile

Example 1: Reddy et al. have synthesized CeO_2–La_2O_3 (8:2 mol ratio) by a modified coprecipitation method using $Ce(NO_3)_3 \cdot 6H_2O$, $La(NO_3)_3 \cdot 6H_2O$ and NH_3 [3]. The sample name was abbreviated as CL calcination temperature (773–1,073 K). The XRD profile of ceria and ceria–lanthana system at different calcination temperature is shown in Figure 6.10. Substitution of Ce^{4+} by lower valent larger size La^{3+} (r_{La3+} = 0.11 nm, r_{Ce4+} = 0.097 nm) should carry lattice expansion (rise of cell parameter), but no appreciable difference in lattice parameter is observed in XRD. However, shifting of diffraction peaks toward lower Bragg's angle side as well as broadening of peak with relatively lower intensity have been observed. Again, with increasing calcination temperature, peaks becomes more crystalline. These observations indicated the incorporation of La^{3+} in ceria lattice without much affecting the lattice cell parameter. Simply, it indicated the formation of ceria–lanthana solid solution which stabilized the ceria lattice after incorporation of La^{3+} against temperature.

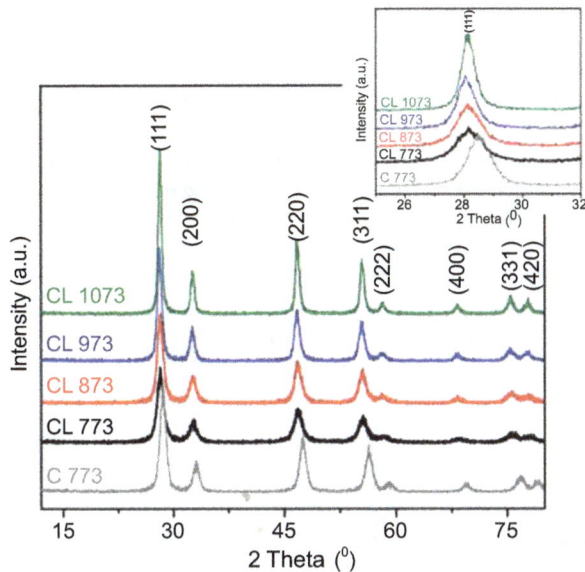

Figure 6.10: The X-ray diffraction profile of ceria (C) and ceria–lanthana (CL 773–1,073) system at different calcination temperature. With permission from American Chemical Society.

Example 2: Rutu et al. have prepared x wt% yttria(100−x) wt% zirconia ($x = 0-20$) support by mechanical mixing. Further, 5 wt% Ni was impregnated over yttria–zirconia support [4]. The catalyst system was employed to dry reforming of methane. The catalyst was abbreviated as Ni–xY–ZrO$_2$ ($x = 5-20$). The XRD profile of ZrO$_2$, Ni–ZrO$_2$, Ni–xY–ZrO$_2$ ($x = 5-20$), spent Ni–ZrO$_2$, spent Ni–xY–ZrO$_2$ ($x = 5-20$) catalyst samples are shown in Figure 6.11. Pure ZrO$_2$ sample had monoclinic zirconia phase. After addition of Ni over ZrO$_2$ (Ni/ZrO$_2$), NiO cubic phase appeared and monoclinic zirconia phase got down and tetragonal zirconia phase was yet to appear. The crystalline size of NiO was 36.8 nm (Figure 6.11A). On addition of Y$_2$O$_3$ in ZrO$_2$ support, Ni-based catalyst system had tetragonal yttrium zirconium oxide phase, as well as cubic yttrium oxide phases appeared. The crystalline size of NiO was also optimized up to 18–22 nm upon yttria addition (Figure 6.11B). The spent ZrO$_2$-supported Ni system (Ni/ZrO$_2$) had metallic Ni phase, graphitic carbon phase and ZrC phase (Figure 6.11C). Possibly during dry reforming reaction over the catalyst surface, CH$_4$ is decomposed into carbon at Ni–Zr interface and the carbon reacted with ZrO$_2$ and ZrC phase was formed. Interestingly, after yttria addition to ZrO$_2$, such Ni-based catalyst system had graphitic carbon phase but no ZrC phase. It indicated that the presence on yttria inhibited the reaction between carbon and ZrO$_2$.

Example 3: Khatri et al. prepared ceria-promoted lanthana–zirconia-supported nickel catalyst by wet impregnation of nickel nitrate precursor and ceria nitrate solutions over lanthana–zirconia [5]. The catalyst system was used for dry reforming of methane. The unpromoted lanthana–zirconia-supported and zirconia-supported catalyst were abbreviated as 5Ni/LaZr and 5Ni/Zr, whereas ceria-promoted catalyst was abbreviated as 5Ni/1CeLaZr. The XRD profile of fresh and spent 5Ni/LaZr, 5Ni/Zr and 5Ni/1CeLaZr system are shown in Figure 6.12. Spent "Ni-supported ZrO$_2$ catalyst" had prominent crystalline carbon deposition as well as monoclinic zirconia phases. On just 1 wt% CeO$_2$ addition, carbon formation minimizes but ZrO$_2$ became defective as well as monoclinic phase changed to cubic zirconia phase (Figure 6.12A). On adding the proportion of lanthana into ZrO$_2$ support, such Ni-based catalyst system had zirconium oxide phase but no lanthanum oxide-based phase (Figure 6.12C, E). However, in spent catalyst system, no defective zirconia and no phase transition were noticed but hexagonal La$_2$O$_3$ phase was formed (Figure 6.12B, D, F). It indicated that presence of lanthana stabilizes the zirconia phases at high temperature reaction. On addition of 1 wt% ceria over lanthana–zirconia-supported Ni catalyst, tetragonal cerium zirconium oxide phase formed over the spent catalyst system (Figure 6.12.B). Ceria promoted spent catalyst had diffused cubic NiO crystalline peak than nonpromoted catalyst. It indicated that ceria addition had also brought high NiO dispersion.

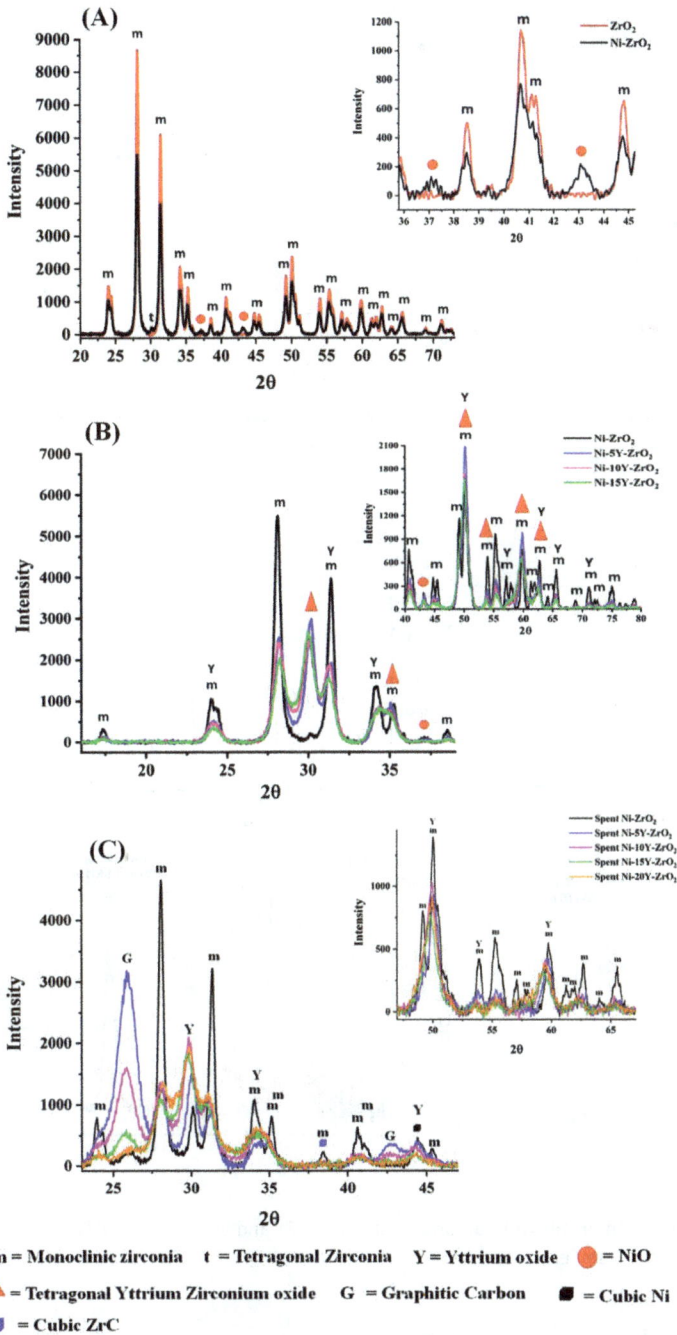

Figure 6.11: The X-ray diffraction profile of ZrO_2, Ni–ZrO_2, Ni–xY–ZrO_2 ($x = 5$–20), spent Ni–ZrO_2, spent Ni–xY–ZrO_2 ($x = 5$–20) catalyst samples. With permission from Elsevier.

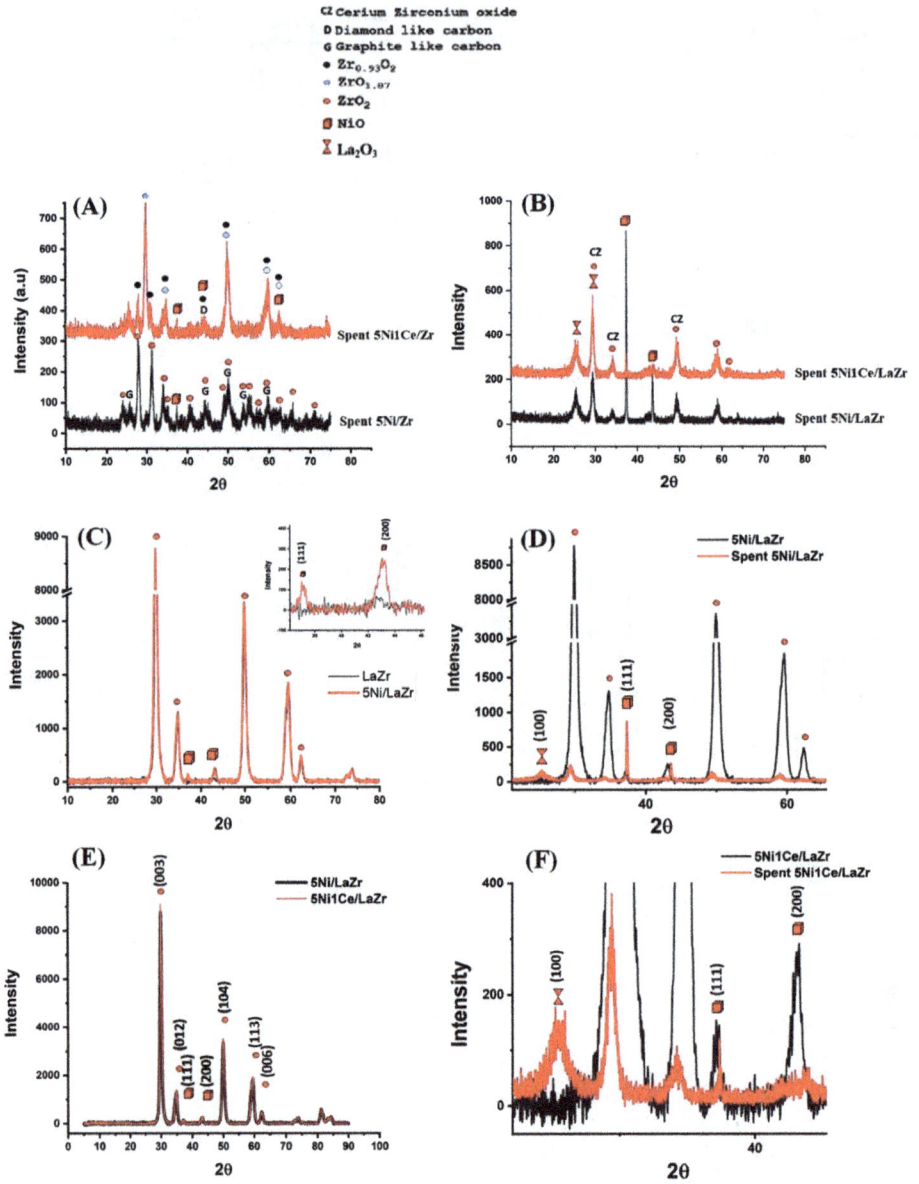

Figure 6.12: The X-ray diffraction profile of fresh and spent 5Ni/LaZr and 5Ni/Zr and 5Ni/1CeLaZr system. With permission from Elsevier.

Example 4: Reddy et al. prepared TiO_2–ZrO_2 mixed oxide (1:1 mole ratio) support by a homogeneous coprecipitation method using $TiCl_4$, $ZrOCl_2$ and urea [6]. Till 600 °C calcination temperature, it was amorphous. At 700 °C, diffraction peaks for $ZrTiO_4$ were clearly noticed and further up to 1,000 °C $ZrTiO_4$ peaks intensity was increased (Figure 6.13; Left). Again, TiO_2–ZrO_2 supported MoO_3 was prepared by wet impregnation method by using ammonium heptamolybdate over TiO_2–ZrO_2 support. On 2% MoO_3 loading, $ZrTiO_4$ as well as $ZrMo_2O_8$ phases were observed at 600 °C calcination temperature. On increasing calcination temperature up to 700 °C, intensity of $ZrMo_2O_8$ phases and TiO_2 phases (anatase and rutile) were grown on the expanse of $ZrTiO_4$ phase (Figure 6.13; right). It indicated that MoO_3 preferably react with zirconia side of $ZrTiO_4$ and formed $ZrMo_2O_8$ with liberating the TiO_2 phase ($ZrTiO_4 + 2MoO_3 \rightarrow ZrMo_2O_8 + TiO_2$).

Example 5: Reddy et al. [7] prepared TiO_2–ZrO_2 mixed oxide (1:1 mole ratio) support as discussed above. Till 500 °C calcination temperature, it was amorphous as observed in XRD. Again, TiO_2–ZrO_2-supported V_2O_5 was prepared by wet impregnation method by using ammonium metavanadate over TiO_2–ZrO_2 support. XRD pattern of TiO_2–ZrO_2-supported 0–100 wt% V_2O_5 at 773 K calcination temperature is shown in Figure 6.14; left and XRD pattern TiO_2–ZrO_2-supported 0–16 wt% V_2O_5 at 973 K calcination temperature is shown in Figure 6.14; right. It was known that 0.07 wt% per m^2 vanadia is needed for monolayer coverage [8] and support TiO_2–ZrO_2 had 160 m^2/g surface area. So, $160 \times 0.07 = 11.2$ wt% vanadia is needed to form monolayer. Now, in XRD, up to 500 °C calcination temperature, TiO_2–ZrO_2-supported V_2O_5 (<12 wt%) was found amorphous but on 12 wt% V_2O_5 load, diffraction peaks V_2O_5 is noticed. By correlating surface area results, V_2O_5 monolayer coverage capacity and XRD, it was concluded that V_2O_5 forms monolayer below 12 wt% loading. Now, if calcination temperature was 700 °C, the support had TiO_2 and ZrO_2 phases whereas as supported V_2O_5 catalyst had prominent ZrV_2O_7 phases and TiO_2 phases. It indicated that V_2O_5 preferably reacted with zirconia side and formed ZrV_2O_7 with liberating the TiO_2 phase ($TiO_2.ZrO_2 + V_2O_5 \rightarrow ZrV_2O_7 + TiO_2$).

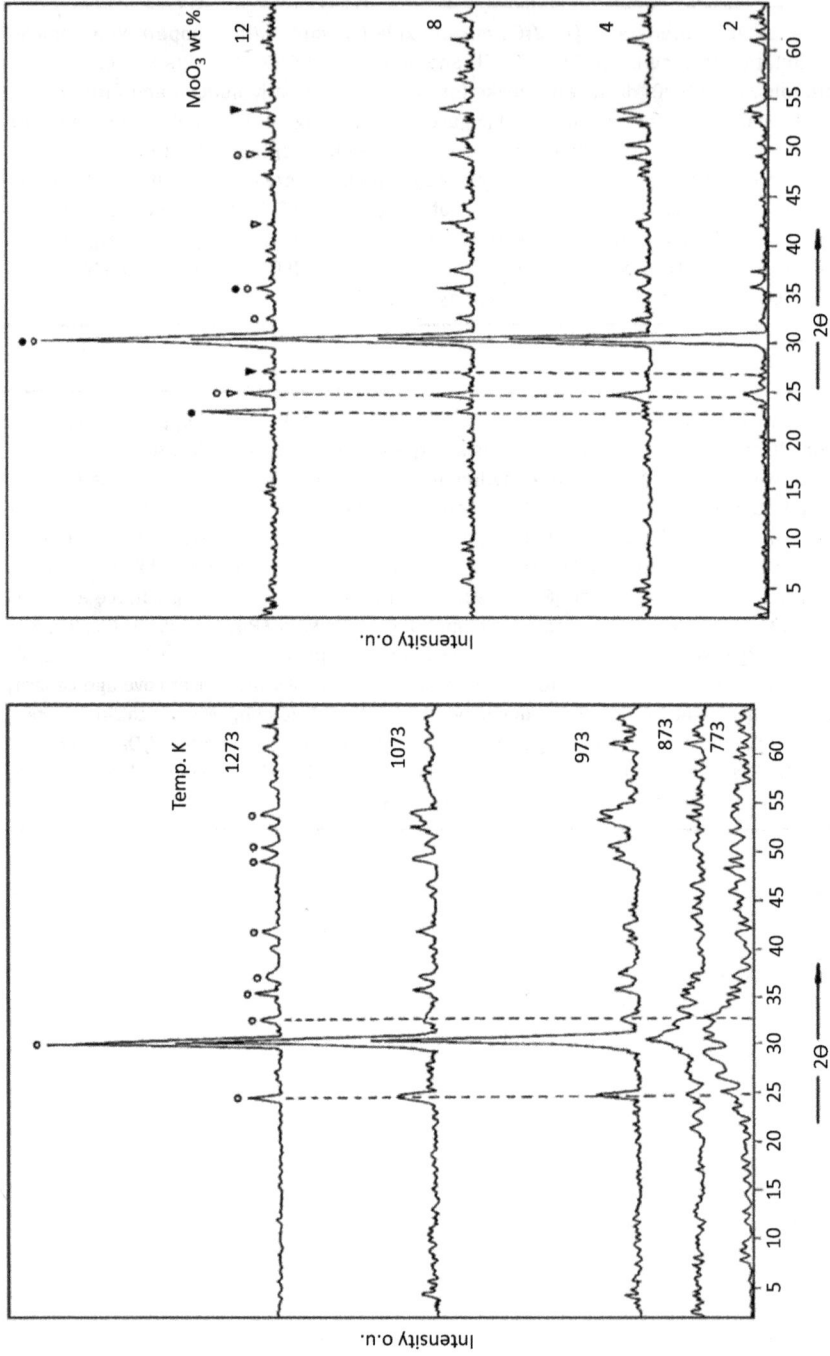

Figure 6.13: Left: XRD pattern of TiO_2–ZrO_2 at different calcination temperatures. **Right:** XRD pattern of TiO_2–ZrO_2-supported 2–12 wt% MoO_3 at 973 K temperature. With permission from Elsevier.

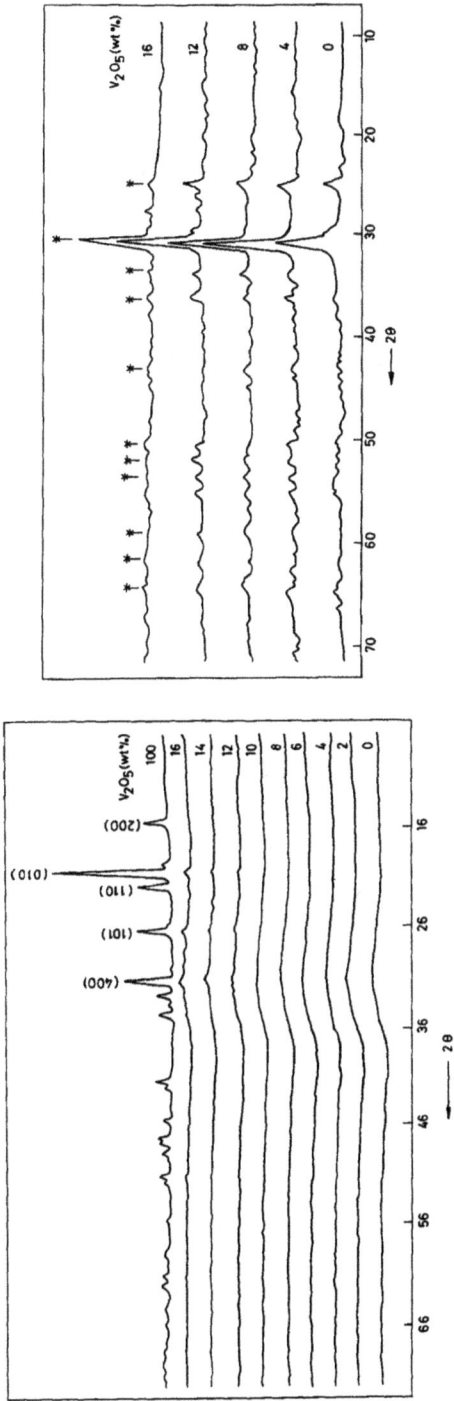

Figure 6.14: Left: XRD pattern of TiO$_2$–ZrO$_2$-supported 0–100 wt% V$_2$O$_5$ at 973 K calcination temperature; **right:** XRD pattern TiO$_2$–ZrO$_2$-supported 0–16 wt% V$_2$O$_5$ at 773 K calcination temperature. With permission from Elsevier.

References

[1]   Mandal, S., Bando, K. K., Santra, C., Maity, S., James, O. O., Mehta, D., Chowdhury, B. *Appl. Catal. A Gen.* **2013**, *452*, 94–104.
[2]   Hassan, S., Kumar, R., Tiwari, A., Song, W., van Haandel, L., Pandey, J. K., Hensen, E., Chowdhury, B. *Mol. Catal.* **2018**, *451* (September 2017), 238–246.
[3]   Reddy, B. M., Katta, L., Thrimurthulu, G. *Chem. Mater.* **2010**, *22* (2), 467–475.
[4]   Patel, R., Fakeeha, A. H., Kasim, S. O., Sofiu, M. L., Ibrahim, A. A., Abasaeed, A. E., Kumar, R., Al-Fatesh, A. S. *Mol. Catal.* **2021**, *510* (April), 111676.
[5]   Khatri, J., Al-Fatesh, A. S., Fakeeh, A. H., Ibrahim, A. A., Abasaeed, A. E., Kasim, S. O., Osman, A. I., Patel, R., Kumar R. *Mol. Catal.* **2021**, *504*, 111498.
[6]   Reddy, B. M., Chowdhury, B. *J. Catal.* **1998**, *179* (2), 413–419.
[7]   Reddy, B. M., Manohar, B., Mehdi. S., J. *Solid State Chem.* **1992**, *97*, 233–238.
[8]   Naga, N. K., Chary, K. V. R., Reddy, B. M., Subrahmanyam. V. S. *Appl. Catal.* **1984**, *9*, 225–233.

7 X-ray photoelectron spectroscopy (XPS)

7.1 Background

In ultrahigh vacuum condition under high temperature operation[1], when an atom is irradiated with X-ray photons, X-ray photons knock the electrons of atom (valance and core electrons) continuously. If X-ray photon energy is sufficiently high, that energy is not only capable of overcoming binding energy of electron from the parent nucleus attraction (E_B) but it can push also electron from atom surface to immediate vacuum (work function, Φ) as well as shoot up the electron (photoelectron) in the vacuum with kinetic energy (E_K):

$$h\upsilon = E_B + \Phi + E_K$$

$$E_B = h\upsilon - (\Phi + E_K)$$

Further, by providing stopping potential (by detector), the kinetic energy of electron is equalized and calculated ($\frac{1}{2}mv_1^2 - \frac{1}{2}mv_o^2 = e(V - V_o)$). The systematic diagram of electron knockout from sample (by X-ray gun) and kinetic energy detection by detector (through stopping potential) is shown in Figure 7.1. By knowing the kinetic energy of shoot up electron or photoelectron, one can calculate the binding energy of that photoelectron in the atom. Binding energy of photoelectrons (valance and core electron) of particular atom is specific. Overall, peculiar binding energy specifies the chemical element composition of sample. The study of interaction of X-rays with matter and thereby the produced photoelectron (due to wave matter interaction) is called X-ray photoelectron spectroscopy (XPS).

7.2 Instrumentation and working principle

The detail of XPS instrumentation along with working principle is discussed one by one in proceeding sections.

7.2.1 Surface preparation by ion gun and then after switching from ion gun to electron gun

Usual analyzing surface has contamination such as adsorbed hydrocarbon, moisture and so on. It is cleaned by argon ion beam treatment (Figure 7.2). Argon ion beam is generated either by electron impact or by gas discharge. Ion gun provides high-energy flux which cleans the surface before examination. High-energy flux of helium ion makes the sample atom sputtered out layer by layer to study the elemental depth profile.

https://doi.org/10.1515/9783110656480-007

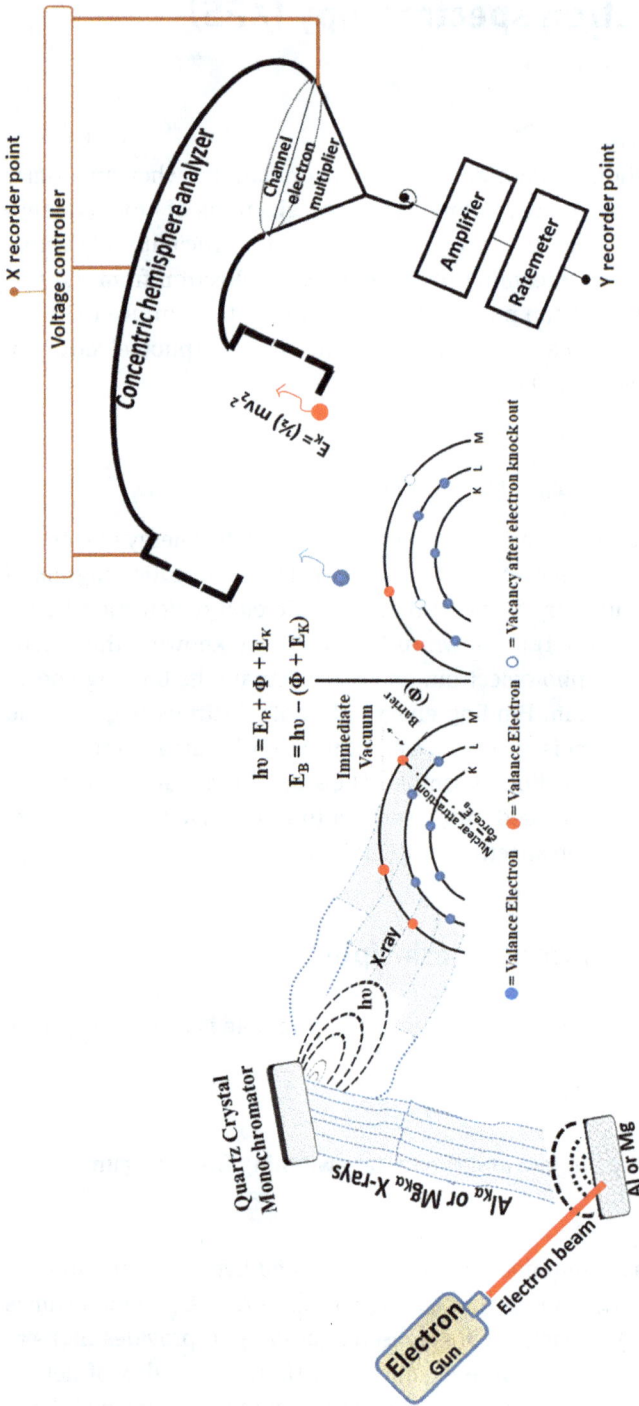

Figure 7.1: Left: The systematic diagram of electron knockout from sample (by X-ray gun), **Right:** kinetic energy detection by detector (through stopping potential).

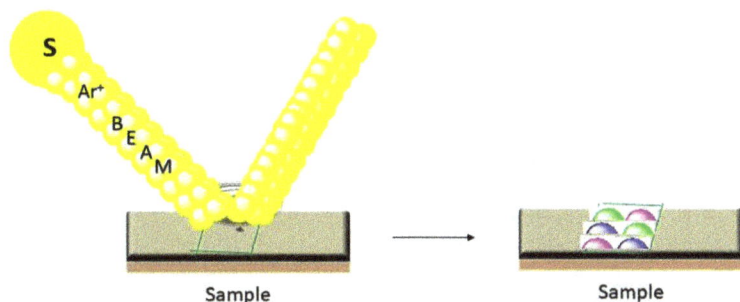

Figure 7.2: Cleaning of sample by argon ion beam before XPS analysis.

After surface preparation, electron gun is switched in place of ion gun. Electron beam is emitted from cathode (through either thermionic or field emission), accelerated by applied voltage (between cathode and anode) and stabilized against voltage fluctuation by Wehnelt electrode (Figure 7.3). Field emission gun provides superior brightness and higher spatial resolution than thermionic LaB_6 gun.

Now, high-energy electrons strike Al (or Mg) metal anode and generate $Al_{k\alpha}$ (or $Mg_{k\alpha}$) X-rays. Electron gun along with striking anode is called X-ray gun. The energy of $Mg_{k\alpha}$ and $Al_{k\alpha}$ X-rays are in range of 1.2536 to 1.4866 eV under the line width of 1 eV. These X-rays are called soft X-rays. To reduce the background intensity, use of monochromator (as quartz crystal) is beneficial. $Al_{k\alpha}$ (or $Mg_{k\alpha}$) X-rays are passed through monochromator. Monochromator allows monochromatic X-rays to glance over the sample

7.2.2 Interaction among X-ray and atom's electron

If X-ray frequency "v" is greater than the threshold frequency of metal "v_o" ($v > v_o$ or $hv > hv_o$), then relative higher energy amount overcomes the binding energy of electron as well as knocks the electron out with certain kinetic energy ($KE = \frac{1}{2} mv^2$). These electrons are called photoelectrons and the study of their spectra is photoelectron spectroscopy. These knocked-out electrons are represented by a symbol having three information. The electron is ejecting (a) from which element? (b) from which orbital? and (c) at what momentum? As electron are ejecting from Al, from 2p orbital and it has 3/2 and ½ total momentum (1 angular momentum +½ spin momentum = 1 ½ = 3/2; same 1–½ = ½), then it is represented as Al $2p_{3/2}$ and Al $2p_{1/2}$ (Figure 7.4 left).

After electron knock out, an atom becomes charged (positive charge) and new event takes place. (Here, photoelectron is ejected from L shell.) Charged atom causes instant pull on outer electron to fill the inner electron vacancy. (Here, vacancy is created at *L shell*.) Electron from higher shell drops to the lower vacant shell position and so emits proportional amount of energy or secondary X-ray (equal to energy

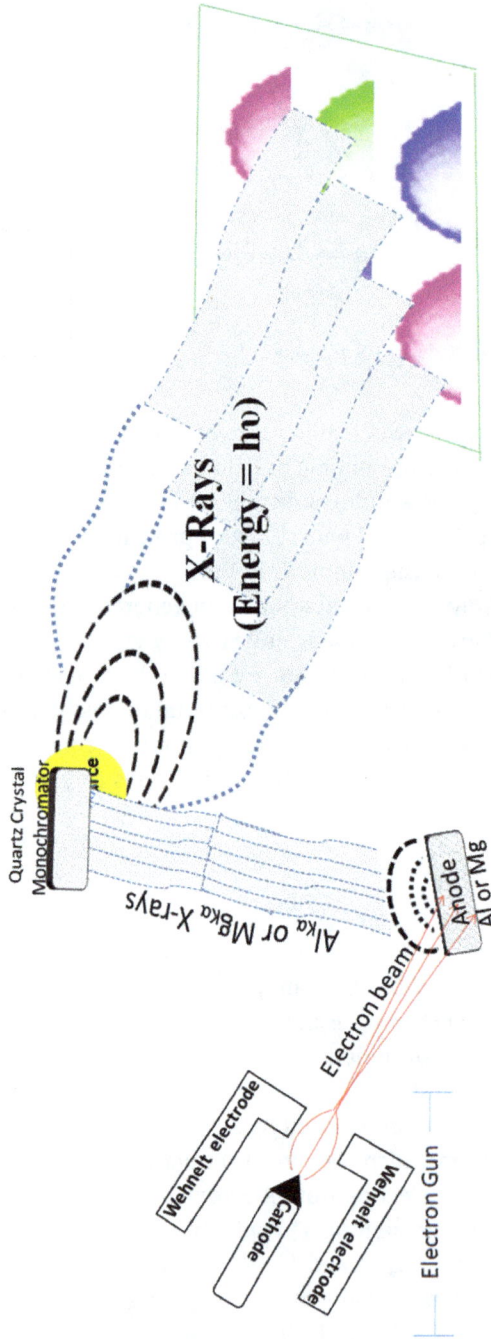

Figure 7.3: Generation of electron beam and X-ray beam.

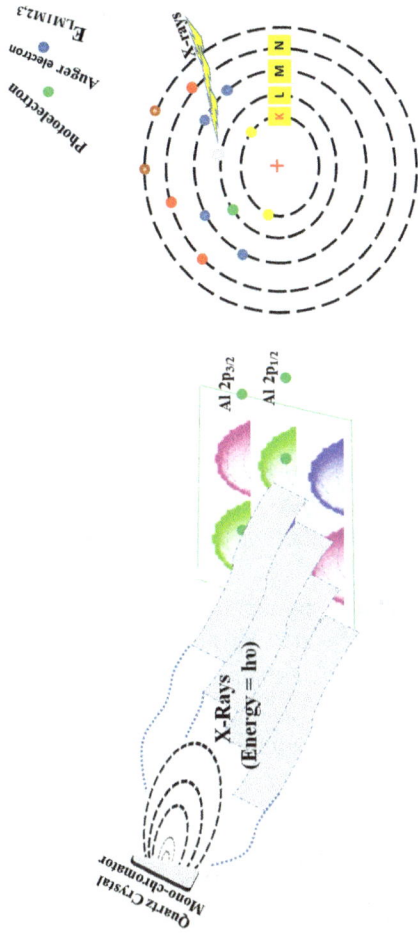

Figure 7.4: Left: Electron knock out from the sample (by X-rays), **Right:** a systematic representation of ejection of "photoelectron and auger electron" from sample, formation of electron vacancy in the atom orbit and generation of secondary X-ray.

difference between shrinking shells). Here, electron from M_1 drops to inner shell. The secondary X-rays may again knock out some electron nearby. Knock-out energy does not only overcome the binding energy of electron but it also puts a kinetic energy to ejected electron. Here, finally, electron from M_2 is emitted. This electron is known as Auger electron photoelectron and study of their spectra is Auger electron spectroscopy. This Auger electron is represented as name of shell having vacancy, dropping shell, shell of ejecting electron. Here, it is $E_{LM1M2,3}$.

7.2.3 Electron energy analysis

The systematic representation of kinetic energy, binding energy measurement and electron intensity measurement of detector are shown in Figure 7.5. Electron energy analyzer measures the kinetic energy of photo- or Auger electron (through retardation against applying electrostatic potential; $\frac{1}{2}mv^2 = eV$) as well as density of that electron. It has electromagnetic lens, concentric hemisphere analyzer (CHA), channel electron multiplier, amplifier and ratemeter.

(1) Electromagnetic lens: Electromagnetic lens retards the kinetic energy of ejected electron (by applied potential as $\frac{1}{2}mv_1^2 - \frac{1}{2}mv_0^2 = e(V - V_0)$) as well focus the electrons on next segment of analyzer (concentric hemisphere) through the slit.

(2) CHA: The section is made up of two concentric hemispheres of radii R_1 and R_2 having negative potentials V_1 and V_2 respectively. These potential generates median equipotential surface V_0 (or pass energy) at radius R_0. Pass energy guides the electron to channel electron multiplier. X recorder point receives data for X-axis of plot. It collects data of potential applied (V) against pass potential (V_0) to retard the kinetic energy of ejected electrons. So, respective kinetic energy value can be collected by equation $\frac{1}{2}mv_1^2 - \frac{1}{2}mv_0^2 = e(V - V_0)$. Thereafter, respective binding energy value for electron is calculated by equation; $E_B = h\upsilon - \emptyset - \frac{1}{2}mv^2$

(3) Channel electron multiplier, amplifier and ratemeter: This setup enables the analyzer to detect the intensity of electron. The data for intensity (for y-axis on plot) is recorded at Y recorder point.

Overall, the X-Y recorder shows the intensity versus binding energy plot for ejected electron. Plot is known as XPS and X-ray Auger electron spectra.

7.2.4 Information from peaks comparison

On comparing two XPS spectra of closely related material, peak growth (lengthening and broadening) and peak shift to higher/lower bonding energy is noticed frequently

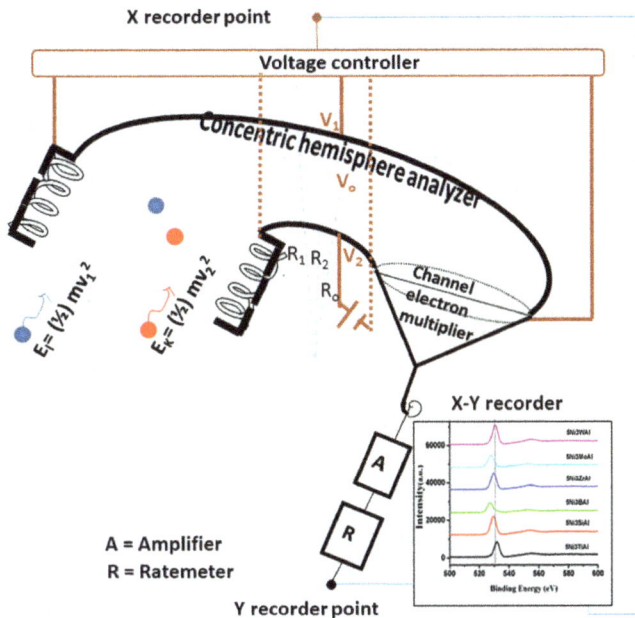

Figure 7.5: The systematic representation of kinetic energy retardation by stopping potential, binding energy measurement by $E_B = hv - (\Phi + E_K)$ and electron intensity measurement by analyzer of detector.

(Figure 7.6). Vertical rise of certain peak (of certain element core spectra) indicates surface enrichment by the particular element. Broadening of certain peak indicates the presence of more than one type of coordination/oxidation state.

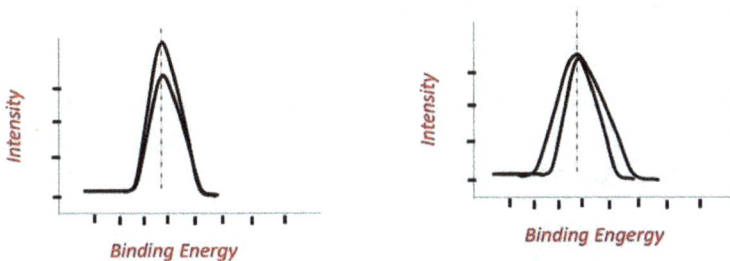

Figure 7.6: Vertical rise and broadening peak of XPS peak.

In XPS, peak shift to higher bonding energy is noticed frequently (intensity versus binding energy plot in Figure 7.7). It may be attributed to atomic dispersion of particular element on the other support oxides (loss of crystallinity) or depletion of negative charge density over particular element. It may be attributed to stronger interaction of particular element (orbital electron) to support. It may be attributed to oxidation of particular element (i.e., Au^0 to Au^{3+}) or increasing ionicity and decreasing covalence. In XPS, peak shift to lower bonding energy is also noticed frequently. It may be attributed to rise of crystallinity or growing of negative charge density over particular element. It may be attributed to inferior interaction of particular element (orbital electron) to support. It may be attributed to reduction of particular element (i.e., Au^{3+} to Au^0) or decreasing ionicity or increasing covalence.

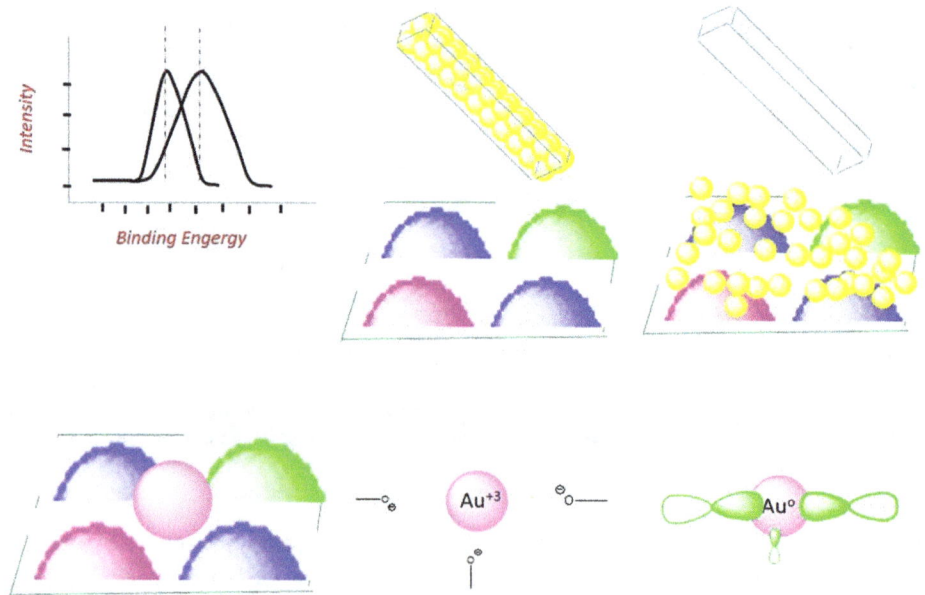

Figure 7.7: A systematic representation of intensity versus binding energy plot, atomic dispersion of particular element on the other support oxides and oxidation of particular element (i.e., Au^0 to Au^{3+}).

With the help of lattice parameters information from XRD results and concentration of different cations by XPS results, the radius of oxygen vacancy (r_{vo}) for a sample can be calculated as per the given formula [1]

$$\frac{\sqrt{3}}{4}(a' - a_0) = \frac{[Ce^{3+}]}{[Ce^{4+}]}\left[r_{Ce^{3+}} - r_{Ce^{4+}} + \frac{1}{4}\left(r_{vo} - r_{O^{2-}} \right) \right]$$

where a' = expanded lattice parameter, a_0 = lattice parameter of CeO_2, $r_{Ce^{3+}}$ = radius of Ce^{3+}, $r_{Ce^{4+}}$ = radius of Ce^{4+}, r_{vo} = radius of oxygen vacancy, $r_{O^{2-}}$ = radius of O^{2-}. concentration of Ce^{3+} can be calculated by XPS Ce (3d) peak area for Ce^{3+} and Ce^{4+} as $[Ce^{3+}] = \frac{\text{Total area for } Ce^{3+}}{\text{Total area for } Ce^{3+} + \text{Total area for } Ce^{4+}} \times 100\%$, $[Ce^{4+}] = 100 - [Ce^{3+}]$

7.3 Glossary

1. Ultrahigh vacuum condition under high temperature operation[1]: XPS instrument is operated under ultrahigh vacuum condition under high-temperature operation so that sample surface may not adsorb monolayer of gas molecule as well as low energy electron (ejecting from sample surface) may not be scattered by gas molecule before reaching the detector. Overall, such instrumentation is arranged for calculating intensity and binding energy of photo/Auger electron.

7.4 Analysis of XPS profile

Example 1: Al-Fatesh et al. synthesized 3 wt% MO_x-doped γ-Al_2O_3 solid solution by mechanically mixing of $(NH_4)_6Mo_7O_{24} \cdot 4H_2O$ or $(NH_4)10H_2(W_2O_7)_6$ precursor salt with γ-Al_2O_3 followed by calcination [2]. Further, 5 wt%Ni was dispersed 3 wt% MO_x-doped γ-Al_2O_3 was prepared by mechanically mixing of nickel nitrate with support followed by calcination. The sample is abbreviated as 5Ni3MAl (M = Mo, Si, B, Zr, W, Ti). The O (1s) XPS spectra of 5Ni3MAl (M = Mo, Si, B, Zr, W, Ti) catalyst system is shown in Figure 7.8. The binding energy value of O (1s) profile 5Ni3MoAl, 5Ni3SiAl, 5Ni3ZrAl and 5Ni3TiAl catalysts (prepared by Al-Fatesh et al. [1]) showed binding energy value at 527.69, 529.30, 529.69 and 531.68 eV, respectively. The O (1 s) XPS binding energy of pure MoO_2, SiO_2, ZrO_2 and TiO_2 were claimed at 530.5, 533.0, 530.4 and 530.1 eV [2]. The binding energy shifts toward lower values than corresponding pure metal oxides indicates an increase in the covalence character of M−O−M' bonds in mixed oxide matrix, which in turn, influenced the formation of nickel aluminum oxide. However, the binding energy shifts toward higher value that of the pure metal oxide in 5Ni3TiAl, indicates the lower covalence character of M−O−M' bonds in mixed oxide matrix which might retard the formation of nickel aluminum oxide.

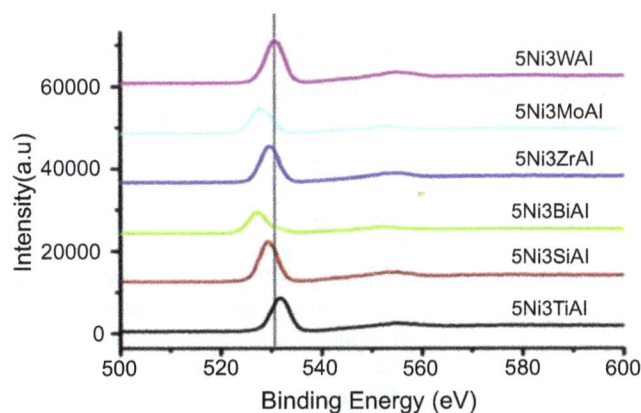

Figure 7.8: The O (1s) XPS spectra of 5Ni3MAl (M = Mo, Si, B, Zr, W, Ti) catalyst system. With permission from Elsevier.

Example 2: Al-Fatesh et al. synthesized 3 wt% MO_x-doped γ-Al_2O_3 solid solution by mechanically mixing of $(NH_4)_6Mo_7O_{24} \cdot 4H_2O$ or $(NH_4)10H_2(W_2O_7)_6$ precursor salt with γ-Al_2O_3 followed by calcination [2]. Further, 5 wt% Ni was dispersed 3 wt% MO_x-doped γ-Al_2O_3 was prepared by mechanically mixing of nickel nitrate with support followed by calcination. The sample is abbreviated as 5Ni3MAl (M = Mo, Si, B, Zr, W, Ti). The Ni(2p) XPS spectra of 5Ni3MAl (M = Mo, Si, B, Zr, W, Ti) is shown in Figure 7.9. The binding energy value of Ni (2p) profile for 3 wt% WO_3-doped γ-Al_2O_3-supported Ni catalyst (5Ni3WAl) was shown at 855.5 eV which was attributed to characteristic peak of Ni^{2+} ($2p_{3/2}$). However, on different promoter, binding energy values were shifted to higher binding energy which indicated the promotor identity and interaction with NiO. The low intensity peak in the case of titanium or boron-doped γ-Al_2O_3-supported Ni catalyst claimed absence of nickel aluminum oxide and presence of only nickel oxide.

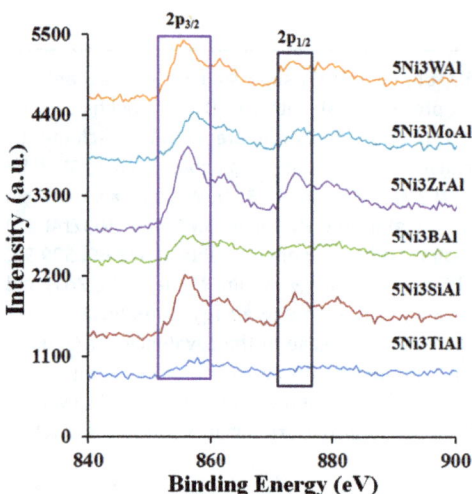

Figure 7.9: The Ni(2p) XPS spectra of 5Ni3MAl (M = Mo, Si, B, Zr, W, Ti). With permission from Elsevier.

Example 3: Mandal et al.[3] have prepared ceria-promoted titanium mesoporous 3D tunable silicates (or Ce–Ti–TUD-1) [3]. The Ti (2p) XPS spectra of TUD-1, 0.1 mol % Ce–Ti–TUD-1 and 0.5 mol % Ce–Ti–TUD-1 samples is shown in Figure 7.10. If electron comes from core p orbital, it has angular momentum is 1 and spin momentum is ½. If it unifies antiparallelly, it gives 1–½ = 1/2 total momentum and lower binding energy value. This peak is shown by Ti($2P_{1/2}$). If it unifies parallelly, it gives 1 + ½ = 3/2 total momentum and lower binding energy value. This peak is shown by Ti($2P_{3/2}$). One peak Ti($2P_{1/2}$) at around 461.1 eV is attributed by Ti species in tetrahedral arrangement. After cerium doping on Ti–TUD-1 material, the peak shifts toward the lower binding energy with some larger peak area. Peak shifts toward the lower binding energy is due to differential charge up property and spatial distribution of tetrahedral titanium species. Larger peak area indicates surface enrichment of Ti^{4+} site.

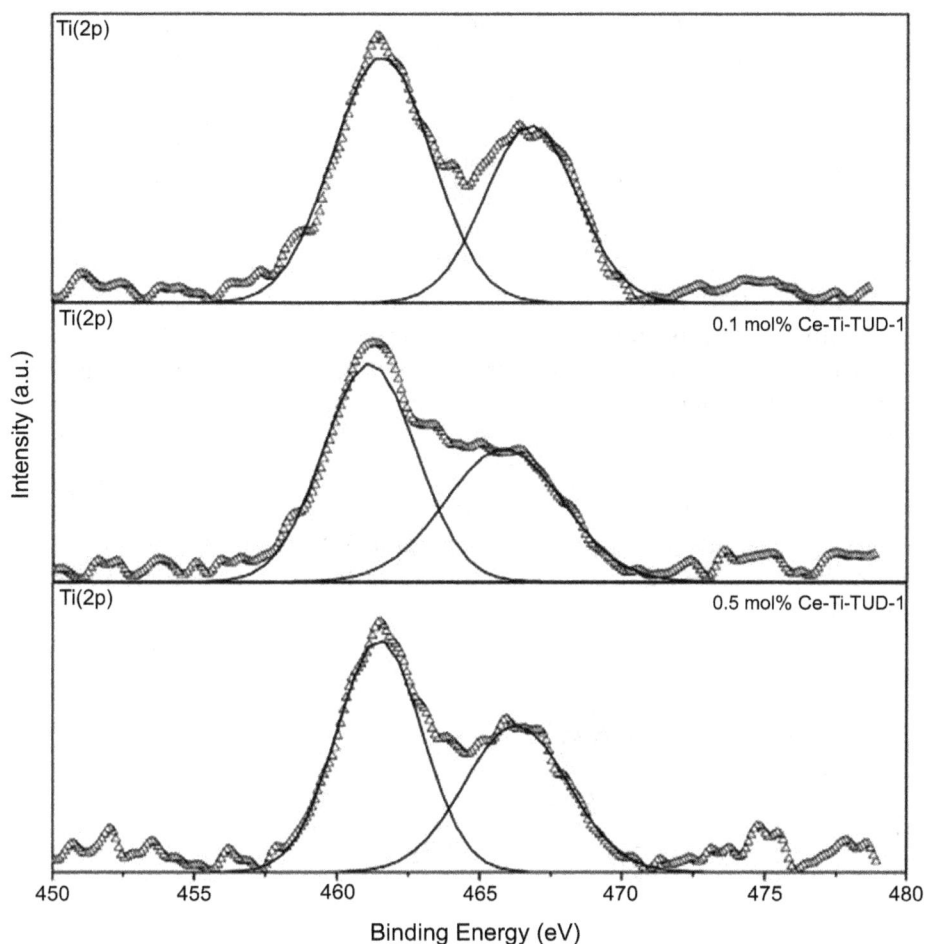

Figure 7.10: The Ti (2p) XPS spectra of TUD-1, 0.1 mol % Ce–Ti–TUD-1 and 0.5 mol % Ce–Ti–TUD-1 samples. With permission from Elsevier.

Example 4: Hassan et al.[4] have prepared cobalt-doped ceria catalyst by the nonhydrothermal sol–gel method. The catalyst is abbreviated as Ce(4)–CeO$_2$ (Ce is 4 mol %). Ce (3d) XPS spectra of CeO$_2$ and Ce(4)–CeO$_2$ samples are shown in Figure 7.11. For understanding Ce (3d) XPS peaks of sample, we have to understand the following facts [4]. First of all, Let us consider Ce^{4+}, the correct electronic configuration of Ce^{4+} is $3d^9 4f^1 5d^0 6s^0$, not $3d^{10} 4f^0 5d^0 6s^0$. O^{2-} has electronic configuration is $1s^2 2s^2 2p^6$. The overall electronic configuration can be shown as $(Ce^{+4}) 3d^9 4f^1 5d^0 6s^0 (O^{2-}) 2p^6$. After radiation, electrons from core 3d orbital of cerium comes out with either total momentum 3/2 (angular momentum–spin momentum = 2–1/2 = 3/2; labeling u''') or total momentum 5/2 (angular momentum + spin momentum = 2 + 1/2 = 5/2; labeling v''').

Again for Ce^{4+}, electron may be transferred from oxygen to the empty orbital of f^0. After this transition, the electronic configuration is modified for one electron transition as $(Ce^{4+})\,3d^9 4f^2 5d^0 6s^0\,(O^{2-})\,2p^5$ and for two electrons as $(Ce^{4+})\,3d^9 4f^3 5d^0 6s^0\,(O^{2-})\,2p^4$. After radiation, electrons from core 3d orbital come out with either total momentum 3/2 (angular momentum–spin momentum = 2–1/2 = 3/2; labeling u'') or total momentum 5/2 (angular momentum + spin momentum = 2 + 1/2 = 5/2; labeling v'').

Now again, let us consider Ce^{3+}, the electronic configuration of Ce^{3+} is $3d^9 4f^2 5d^0 6s^0$. O^{2-} has electronic configuration is $1s^2 2s^2 2p^6$. The overall electronic configuration can be shown as $(Ce^{4+})\,3d^9 4f^2 5d^0 6s^0\,(O^{2-})\,2p^6$. If this structure is radiated by X-ray, then electrons from core 3d orbital of cerium come out with either total momentum 3/2 (angular momentum–spin momentum = 2–1/2 = 3/2; labeling u') or total momentum 5/2 (angular momentum + spin momentum = 2 + 1/2 = 5/2; labeling v'). Again, for Ce^{3+} after considering hybridization, electron transfer from oxygen to the empty orbital of f^2 can be considered. After radiation, electrons from core 3d orbital come out with either total momentum 3/2 (angular momentum–spin momentum = 2–1/2 = 3/2; labeling u_o) or total momentum 5/2 (angular momentum + spin momentum = 2 + 1./2 = 5/2; labeling v_o). Overall surface is composed of a mixture of Ce^{3+}/Ce^{4+} oxidation states.

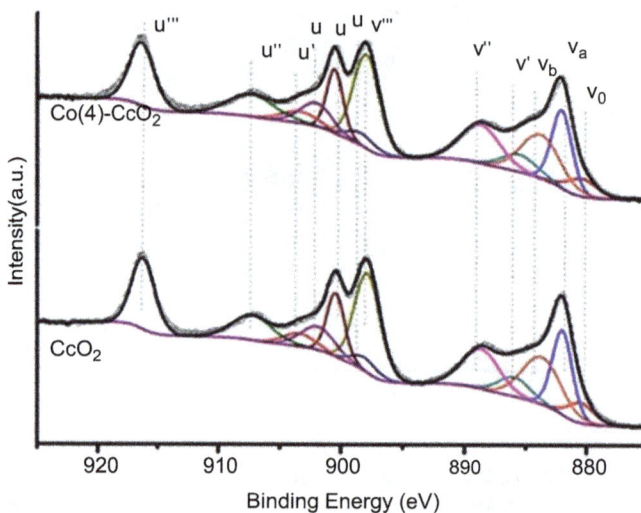

Figure 7.11: Ce (3d) XPS spectra of CeO_2 and Ce(4)–CeO_2 samples. With permission from Elsevier.

Example 5: Rutu et al. have prepared $x$wt% yttria (100–x)wt% zirconia (x = 0–20) support by mechanical mixing. Further, 5 wt% Ni was impregnated over yttria–zirconia support [5]. The catalyst was abbreviated as Ni–xY–ZrO$_2$ (x = 0–20). The O (1s) XPS spectra of Ni–ZrO$_2$, Ni–15Y–ZrO$_2$ and spent Ni–15Y–ZrO$_2$ samples are shown in Figure 7.12. The effect of yttria addition in support was clearly observed in O (1s) XPS spectra of zirconia supported Ni catalyst and yttria–zirconia-supported Ni catalyst. Zirconia-supported Ni catalyst has O (1s) XPS peaks at 529.3 eV for lattice oxygen and 532.4 eV for oxygen in adsorbed water whereas 15 wt% yttria–85 wt% zirconia-supported Ni catalyst had addition peak at 537.7 eV for oxygen (O$_2$) coverage at surface. The catalyst system was employed for dry reforming of methane. Interestingly, in spent catalyst system, peaks due to lattice oxygen as well as "oxygen coverage at surface" were lost due to possible utilization of these oxide species in carbon deposit oxidation during dry reforming reaction. It clearly indicated the enrichment of catalyst surface by adsorbed oxygen in yttria–zirconia-supported nickel catalyst system.

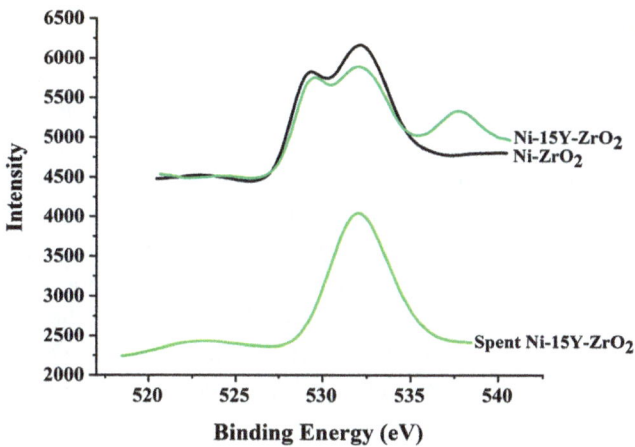

Figure 7.12: The O (1s) XPS spectra of Ni–ZrO$_2$, Ni–15Y–ZrO$_2$ and spent Ni–15Y–ZrO$_2$ samples. With permission from Elsevier.

References

[1]   Zhang, G., Shen, Z., Liu, M., Guo, C., Sun, P., Yuan, Z., Li, B., Ding, D., Chen, T. *J. Phys. Chem. B* **2006**, *110* (51), 25782–25790.

[2]   Al-Fatesh, A. S., Kumar, R., Kasim, S. O., Ibrahim, A. A., Fakeeha, A. H., Abasaeed, A. E., Alrasheed, R., Bagabas, A., Chaudhary, M. L., Frusteri, F., Chowdhury, B. *Catal. Today* **2020**, *348* (September), 236–242.

[3]   Mandal, S., Rahman, S., Kumar, R., Bando, K. K., Chowdhury, B. *Catal. Commun.* **2014**, *46*, 123–127.

[4]   Hassan, S., Kumar, R., Tiwari, A., Song, W., van Haandel, L., Pandey, J. K., Hensen, E., Chowdhury, B. *Mol. Catal.* **2018**, *451* (September 2017), 238–246.

[5]   Patel, R., Fakeeha, A. H., Kasim, S. O., Sofiu, M. L., Ibrahim, A. A., Abasaeed, A. E., Kumar, R., Al-Fatesh, A. S. *Mol. Catal.* **2021**, *510* (April), 111676.

8 Scanning electron microscopy (SEM)

8.1 Background

Examination of ultrafine microstructures by focus electron beam scanning the whole surface of specimen is known as scanning electron microscopy and the instrument is scanning electron microscope (SEM). Because of high resolution and large depth of field, SEM image has three-dimensional appearances. SEM instrument can be equipped with additional X-ray energy dispersive spectrometer. SEM is able to provide 20–100,000 times image magnification.

8.2 Instrumentation and working principle

The systematic schematic diagram of scanning electron microscopy is shown in Figure 8.1. The instrumental set up for SEM is very similar to TEM except less acceleration voltage for electron beam in the range of 1–40 kV (which is one order less than TEM) and electron beam is condensed to a fine probe for surface scanning. Two electromagnetic condenser lenses reduce the crossover diameter of electron beam and then objective electromagnetic lens focus the condensed beam as a probe with a nanometer scale diameter. Further, electron beam is guided to specimen where it interacts with the specimen. The beam deflection system[a] is incorporated within the objective which moves the probe over specimen surface along a line and then displaces the probe to the next line.

During scanning, high-energy electrons from probe strike the specimen[b] where some are absorbed by the specimen atoms and some are deflected by the specimen atoms. Specimen atom absorbs electrons and produces equivalent numbers of electron with less velocity/energy at small angle. This scattering is called inelastic scattering and scattered electrons are secondary electrons (SE). These SE are generated in whole sample but it can escape only from a volume near the specimen surface with a depth 5–50 nm. As per atomic mass of atom in the specimen, electrons are also scattered elastically with same velocity/energy as incident electron at larger angle. This scattering is called elastic scattering and scattered electrons are backscattered electrons (BSE). Their high energy enables them to escape from more depth of 50–300 nm. SEs travel larger distance and reach at Everhart–Thornley detector (E-T detector) while BSE travel directly toward detector. The detector is composed of Faraday cage, scintillator, light guide and photomultiplier tube. Faraday cage is either positively or negatively charged. When it is positively charged, low-energy SE is attracted, and when it is negatively charged, low-energy SE (less than 50 eV) is screened out. So, the first case is ideal for analysis of SEs and the second case is for analysis of BSEs.

https://doi.org/10.1515/9783110656480-008

The electron from the specimen surface facing the detector will be collected abundantly causing bright image pattern whereas electrons from the surface not facing the detector will reach the detector difficultly causing darker image pattern. In this way, different region of surface (as per the orientation toward detector) is viewed as brighter to darker pattern on the display screen. This effect is known as trajectory effect. When electrons hit the surface at angles (over edges and cavities), more SEs are emitted than the flat surface. So, these regions also appeared as brighter contrast on the display screen. This effect is known as electron number effect. Collection of electrons from surface at detector due to trajectory effect and electron number effect contribute topographic contrast. For topographic contrast, SE are primary signals.

When the Faraday cage is negatively charged, the BSE signals are chiefly collected at detector. The backscatter coefficient (η) is defined as ratio of BSE escaping from the specimen to number of incident electron. It is found that backscatter coefficient (η) increases monotonically with the atomic number. That means, at the detector, more BSE signal is coming from the specimen regions having higher atomic masses forming brighter pattern on display screen. Similarly, an area with lighter atomic masses will appear in grey to dark pattern on display screen. So, as per the atomic mass composition of total scanning area specimen, brighter to darker pattern will be displayed. This contrast is known as atomic number contrast and compositional contrast.

Scintillator accelerates the respective electron with 12 kV and strikes them over a disk (8–20 mm diameter) to convert electron signal into photon. These photons travel through light guide and photomultiplier tube for 10^6 times signal gain. The output signal is further amplified for display on screen under digital imaging.[c] It should be noted that fluctuation of electron beam current and signal amplification in the detector cannot be totally eliminated. But it constitutes the background noise[d] which causes image obscurity. For getting high-resolution image apart from reducing probe diameter,[e] optimum probe current[f] is needed so that sufficient electrons signal generate at detector which can overcome the background noise. Due to excellent depth of field[g] (which solely depends on working distances and aperture size), SEM showed many times sharper image.

8.3 Glossary

[a]**Beam deflection system:** The beam deflection system has two pair of electromagnetic coil. The first one bends the beam off the optical axis and second one bends the beam back onto the optical axis at pivot point of scan.

[b]**Specimen:** The specimen preparation in SEM is simpler than TEM. It only needs properly dehydrated, contamination free, conductive film coated and fine-sized (about a 10th of the centimeter as specification of specimen holder) specimen. Electron beam heats the specimen water and cause it to burst through the specimen

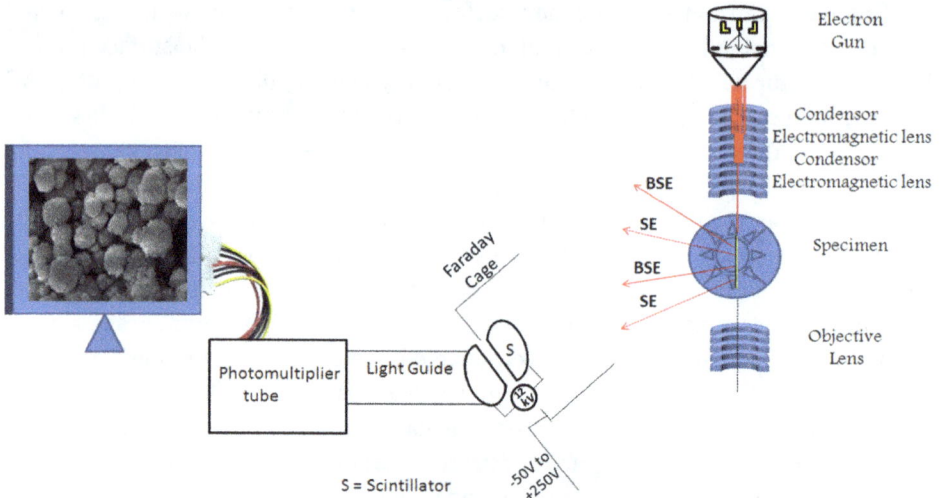

Figure 8.1: The systematic schematic diagram of scanning electron microscopy. With permission from Elsevier.

surface and thus destroy the specimen morphology. So, a properly dehydrated sample (critical-point drying and freeze drying in case of living cell) is needed. Electron beam decomposes hydrocarbon (contamination) and leaves carbon deposit which generates a dark rectangular mark (artifact) in SEM image. Generally, electrons from electron beams are accumulating in nonconductive regions of specimen causing deflection of coming electron beam in an irregular manner. It results into image distortion and artifacts. To prevent the charging, a conductive film coating (through vacuum evaporation or sputtering) over specimen is needed.

[c]**Digital imaging:** The signal electrons emitted from the specimen are collected by the detector, amplified and used to reconstruct the image according to one-to-one correlation between scanning points on the specimen and picture points. Smallest addressable picture element in the display device is pixel. During the scanning, probe moves a rectangular area (specimen raster) on the specimen and the corresponding rectangular image as grid of pixel[b] (image raster) is formed on display screen (Figure 8.2). Raster displayed on the display screen is larger than the corresponding raster scanned by the electron beam on the specimen. Magnification in SEM is defined as ratio of the linear size of the display screen to the linear size of specimen area under scanning.

Figure 8.2: Image raster on screen vis-à-vis specimen raster on specimen.

Overall, raster scanning causes the beam to sequentially cover a rectangular area on the specimen or image. By collecting all image raster, the image of specimen is displayed.

[d]**Noise:** Due to electron current fluctuation and signal amplification in detector, noise has been generated. If image signal is represented by curve line varying with scan position, background noise is represented by vertical fluctuation imposed on curved line. Noise creates image obscurity.

If "n" is the count of signal at detector, noise is square root of "n" counts of signal ($N = (\bar{n})^{1/2}$) at detector. Rose visibility criterion said that to view two distinct object, change in signal (ΔS) (due to object contrast) should be at least five times larger than noise (N).

$$\Delta S > 5N$$

$$\Delta S > 5(\bar{n})^{1/2}$$

ΔS/S is contrast

$$\frac{\Delta S}{S} > \frac{5(\bar{n})^{1/2}}{\bar{n}}$$

$$\frac{\Delta S}{S} = C > \frac{5(\bar{n})^{1/2}}{\bar{n}}$$

[e]**Probe diameter:** probe diameter d_p is approximately expressed as

$$d_p = \frac{2}{\pi\alpha}\left(\frac{i_p}{\beta}\right)$$

α is convergence angle of probe, β is beam brightness, i_p is probe current. Brightness depends on type of electron source used as well as acceleration voltage (or less wavelength). A field emission gun is 1,000 times brighter than a tungsten thermionic gun and 100 times than LaB_6.

From the above equation, it can be said that to minimize the probe diameter convergence angle, beam brightness has to be increased. A limitation that convergence angle has is that at higher angle spherical aberration pronounced affectively. So, convergence angle can be optimized. After optimization, minimum probe size is expressed as

$$d_{min} = KC_s^{1/4}\left(\frac{i_p}{\beta} + \lambda^2\right)^{3/8}$$

K is constant, C_s is spherical aberration coefficient and λ is wavelength of electrons.

Now, conclusion can be drawn that to attain the minimal probe size apart from selecting electron gun, one should decrease wavelength, spherical aberration and increase in brightness of electron illumination.

fProbe current: The probe current (i_p) is proportional to the current of signal electron (i_s) at detector or $i_p = i_s/\varepsilon$; where ε is the proportionality factor.

The current of signal electron (i_s) is

$$i_s = \bar{n}/t; \tag{8.1}$$

$\bar{n}$ can be calculated by the Rose visibility criterion:

$$\frac{\Delta S}{S} = C > \frac{5(\bar{n})^{1/2}}{\bar{n}}$$

$$\bar{n} > \left(\frac{5}{C}\right)^2 \tag{8.2}$$

Putting the value of $\bar{n}$, signal current (i_s) can be calculated as

$$i_s > \frac{25e}{C^2 t}$$

But probe current (i_p) is proportional to signal current:

$$i_p > \frac{25e}{\varepsilon C^2 t}$$

"ε" is proportionality factor between probe current and signal current. "t" is dwell time of probe on object. Dwell time is the time needed to scan (one frame) per number of pixel (in one frame) or ($t = t_f/n_{PE}$).

$$i_p > \frac{25 e n_{PE}}{\varepsilon C^2 t_f}$$

gDepth of field (D_f): Depth of field is tolerance of object position to get the sharper image. It refers to the range of position for an object in which image sharpness does not change. The schematic diagram of depth of field is presented in Figure 8.3.

$$D_f = 2R/\tan\alpha$$

R is resolution, α is angle of convergence or maximum angle of electron beam for exit and enters into a lens. For very small angle, $\tan\alpha \sim \alpha$.

Product of resolution and magnification is called pixel size. If pixel size is 100 μm, then resolution will be $100/M$ μm (pixel $= R$ x M; $R =$ pixel$/M = 100/M$) and depth of field is

$$D_f = 200/(\alpha M)\,\mu\,m$$

By geometry, α is ratio of aperture radius (R_{ape}) working distance (D_w)

$$D_f = (200\,D_w)/(\,R_{ape}M)\,\mu\,m$$

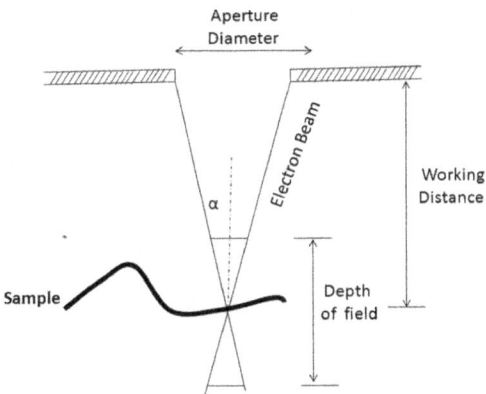

Figure 8.3: The schematic diagram of depth of field.

Clearly small aperture size and long working distance result in high depth of field. However, small aperture size may reduce the probe current and signal to noise ratio may reduce. On fix aperture size, long working distance causes small angle of convergence (α) and resolution will be affected. Additionally, long working distance increases spherical aberration.

8.4 Analysis of SEM image

Example 1: Sumbul et al. prepared Co-incorporated hexagonally ordered periodic mesoporous silica by sol–gel method by using the following gel composition: pluronic P123:H_2O:HCl:n-butanol: tetraethyl orthosilicate:Co(NO$_3$)$_2$ 6H_2O:Fe(NO$_3$)$_2$ = 0.017:200:5.4:1.325:1:0.05–0.15:0.0025 [1]. The catalyst is abbreviated as Fe–Co–HMS–X. The SEM image of Fe–Co–HMS–X is shown in Figure 8.4. The properly dried and calcined (at 540 °C for 24 h) material having Co/Fe = 20 is analyzed by SEM. The material showed wormlike morphology with minimum aggregation.

Figure 8.4: The SEM image of Fe–Co–HMS–X. With permission from Elsevier.

Example 2: Sumbul et al. prepared Nb-incorporated hexagonally ordered periodic mesoporous silica by sol–gel method by using the following gel composition: pluronic P123:H_2O:HCl:n-butanol: tetraethyl orthosilicate:$NbCl_5$ = 0.017:200:5.4:1.325:1:0.075 [2]. The catalyst is abbreviated as Nb–HMS–X (Si/Nb = 13). The field emission SEM image of Nb–HMS–X (Si/Nb = 13) in different conditions is shown in Figure 8.5. The irregular morphology is observed for as-synthesized material Figure 8.5(b). On applying 27 °C synthesis temperature, 100 °C hydrothermal temperature and 540 °C calcination temperature, morphology was disordered Figure 8.5(c). On increasing synthesis temperature to 35 °C, spherical morphology with particle size 500 nm was observed Figure 8.5(a). On further increment of synthesis temperature to 50 °C, rod like morphology (1 µm length and 300 nm diameter) was observed (Figure 8.5(d)). On decreasing the duration of hydrothermal treatment from 24 to 18 h, tube sphere like morphology was appeared (Figure 8.5(f)). On decreasing the hydrothermal treatment's duration further (12 h) (Figure 8.5€), rod like morphology (as like as Figure 8.5(d)) was observed. On decreasing the hydrothermal temperature to 80 °C, bigger particle 600 nm was observed (Figure 8.5(g)). On increasing the hydrothermal temperature from 100 to 120 °C, morphology changed from spherical to tube.

Example 3: Rawesh et al. [3] prepared indium oxide incorporated into three-dimensional spongelike mesoporous silicates by sol–gel technique by using following gel composition: tetraethyl orthosilicate:$In(NO_3)_3$:triethanolamine:H_2O:tetraethylammonium hydroxide = 1:0.08:2:11:1. The catalyst is abbreviated as InO_x/TUD-1; (In/Si = 8/100). After drying and calcining at 700 °C for 10 h, the material showed sponge like morphology (Figure 8.6).

Example 4: Mandal et al. [4] prepared indium oxide incorporated into three-dimensional spongelike mesoporous silicates (Ga–TUD-1) by sol–gel technique by using following gel composition: tetraethyl orthosilicate:gallium nitrate:triethanolamine:tetraethylammonium hydroxide:H_2O = 1:0. 01–0.04:3.0:0.3:15. After drying and calcining at 700 °C for 10 h, the material showed cubic morphology (Figure 8.7).

Example 5: Al-Fatesh et al. synthesized 3 wt% MO_x-doped $γ$-Al_2O_3 solid solution by mechanically mixing of $(NH_4)_6Mo_7O_{24} \cdot 4H_2O$ or $(NH_4)10H_2(W_2O_7)_6$ precursor salt with $γ$-Al_2O_3 followed by calcination [5]. Further, 5 wt% Ni was dispersed 3 wt% MO_x-doped $γ$-Al_2O_3 was prepared by mechanically mixing of nickel nitrate with support followed by calcination. Furthermore, catalyst is employed from dry reforming of reaction. The sample is abbreviated as 5Ni3MAl (M = Mo, W). The SEM of 5Ni3MAl (M = Mo, W) catalyst system is shown in Figure 8.8. After the reaction, SEM of spent catalyst was taken which shows the presence of multiwalled carbon nanotubes on the surface of both catalysts.

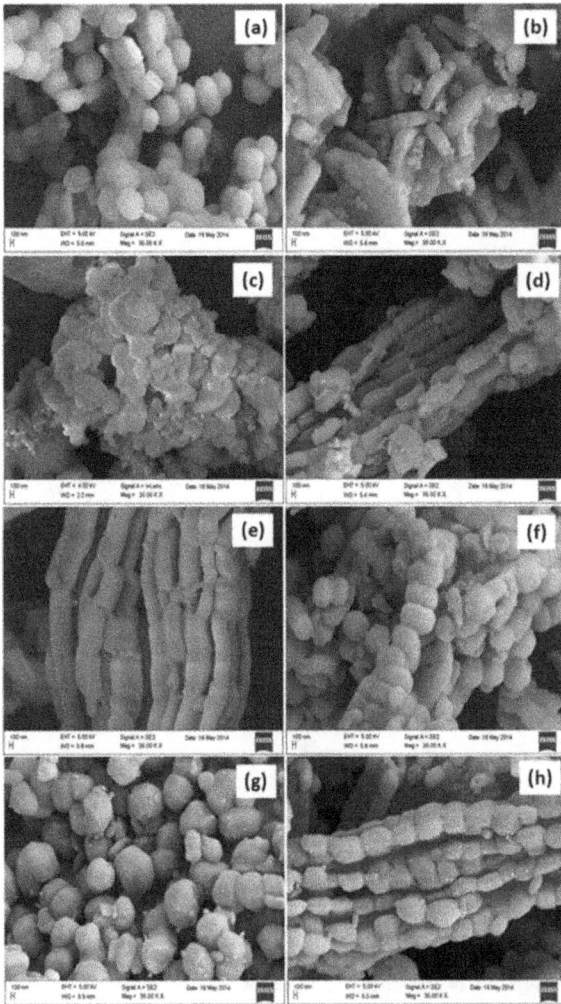

Figure 8.5: The Field emission SEM image of Nb–HMS–X (Si/Nb = 13) in different condition (a) 35 °C synthesis temperature, 100 °C hydrothermal temperature, 540 °C calcination temperature, hydrothermal treatment 24 h, (b) 35 °C synthesis temperature, 100 °C hydrothermal temperature, hydrothermal treatment 24 h (called **as-synthesized material**), (c) **27 °C synthesis temperature,** 100 °C hydrothermal temperature, 540 °C calcination temperature and hydrothermal treatment 24 h, (d) **50 °C synthesis temperature,** 100 °C hydrothermal temperature, 540 °C calcination temperature and hydrothermal treatment 24 h, (e) 35 °C synthesis temperature,100 °C hydrothermal temperature, 540 °C calcination temperature and **hydrothermal treatment 12 h,** (f) 35 °C synthesis temperature, 100 °C hydrothermal temperature, 540 °C calcination temperature and **hydrothermal treatment 18 h,** (g) 35 °C synthesis temperature, **80 °C hydrothermal temperature,** 540 °C calcination temperature and hydrothermal treatment 24 h and (h) 35 °C synthesis temperature, **120 °C hydrothermal temperature,** 540 °C calcination temperature and hydrothermal treatment 24 h. With permission from Elsevier.

Figure 8.6: The field emission SEM image of InO$_x$/TUD-1; (In/Si = 8/100). With permission from Elsevier.

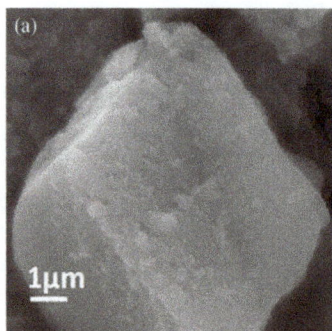

Figure 8.7: SEM image of Ga-TUD-1. With permission from Elsevier.

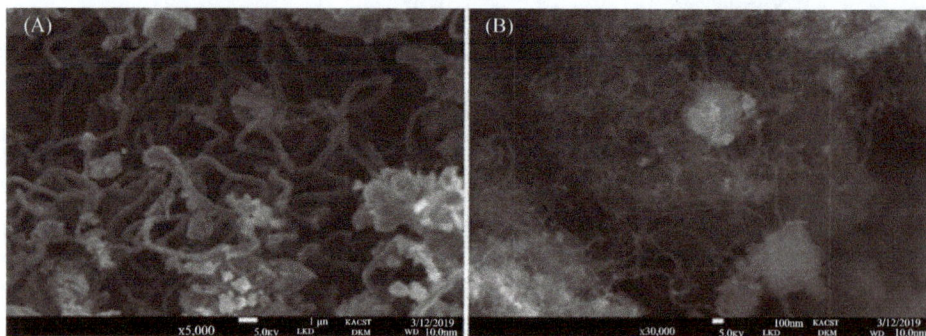

Figure 8.8: The SEM of (A) 5Ni3WAl and (B) 5Ni3MoAl catalyst system. With permission from Elsevier.

References

[1]   Rahman, S., Enjamuri, N., Gomes, R., Bhaumik, A., Sen, D., Pandey, J. K., Mazumdar, S., Chowdhury, B. *Appl. Catal. A Gen.* **2015**, *505*, 515–523.
[2]   Rahman, S., Shah, S., Santra, C., Sen, D., Sharma, S., Pandey, J. K., Mazumder, S., Chowdhury, B. *Micropor. Mesopor. Mater.* **2016**, *226*(2016), 169–178.
[3]   Kumar, R., Das, P. P., Al-Fatesh, A. S., Fakeeha, A. H., Pandey, J. K., Chowdhury, B. *Catal. Commun.* **2016**, *74*, 80–84.
[4]   Mandal, S., Sinhamahapatra, A., Rakesh, B., Kumar, R., Panda, A., Chowdhury, B. *Catal. Commun.* **2011**, *12*(8), 734–738.
[5]   Al-Fatesh, A. S., Kumar, R., Kasim, S. O., Ibrahim, A. A., Fakeeha, A. H., Abasaeed, A. E., Alrasheed, R., Bagabas, A., Chaudhary, M. L., Frusteri, F., Chowdhury, B. *Catal. Today* **2020**, *348*(September), 236–242.

9 Transmission electron microscopy (TEM)

9.1 Background

Electron has wavelength 10,000 times shorter than photons. So, it can be used to examine ultrafine details of microstructure with higher magnification and resolution[a] (up to 0.1 nm) than light microscope. Examination of ultrafine microstructures under instant illumination of electron is known as transmission electron microscopy and the instrument is transmission electron microscope (TEM). TEM is able to provide 500–1,000,000 times image magnification.[b]

9.2 Instrumentation and working principle

TEM has a set of electron gun, specimen tray, series of electromagnetic lens and screen. Electron gun is composed of cathode, Wehnelt electrode and anode. The systematic diagram of electron gun is shown in Figure 9.1. Anode is biased at high voltage (generally 100 kV) with respect to cathode, whereas the Wehnelt electrode is negatively biased with few hundred volts with respect to cathode. Electron beam is emitted from cathode (through either thermionic[c] or field emission[d]), accelerated by applied voltage (between cathode and anode) and stabilized against voltage fluctuation by theWehnelt electrode[e]. A repulsive ring placed between anode and cathode focuses the electron beam onto a small hole at the anode so that the electrons can pass through the anode.

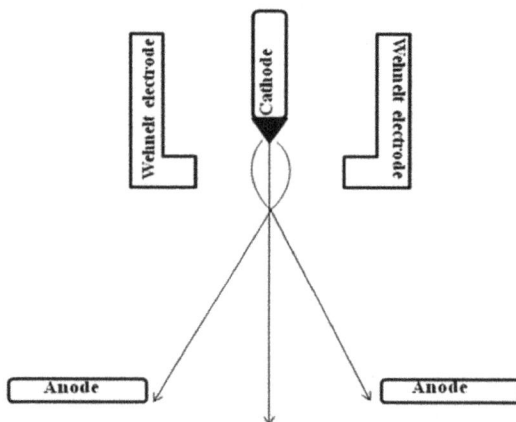

Figure 9.1: A systematic diagram of electron gun (composed of cathode, the Wehnelt electrode and anode).

https://doi.org/10.1515/9783110656480-009

Electron beam is condensed by condenser electromagnetic lens[f]. Now, electron beam interacted with thin dried carbon-coated specimen[g] in the 3-mm mesh disc (made up of copper). After interaction, most of the beams are transmitted and some beams which satisfy the Bragg's angle and crystal orientation of specimen become the part constructive diffraction. On tilting the specimen, crystallographic planes change and so diffraction pattern changes. The transmitted beam is parallel to the incident beam whereas diffracted beam is deflected by some angle (≤1o). A path difference "ΔX" is created between reflected beam on successive plane (Figure 9.2 **Left**). Interplanar distance "d" is related to path difference "ΔX" and glancing angle "θ" is related to Bragg's equation; $\Delta X = 2d \sin\theta$ (Figure 9.2 **Right**). Again, the path difference "ΔX" between reflected rays (from successive plane) is equal to $n\lambda$. So, the relation between electron wave length, interplanar distance and glancing angle is given by equation $n\lambda = 2d \sin\theta$. But for TEM, $\theta < 1$o, then $\sin\theta \sim \theta$. Then, the equation becomes $n\lambda = 2d\,\theta$.

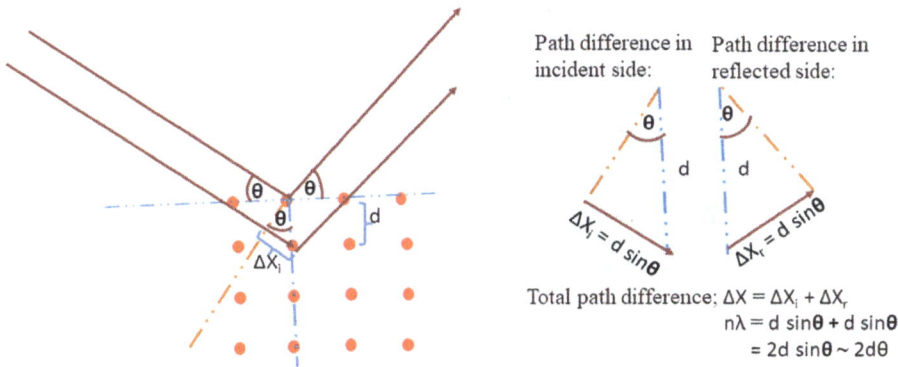

Path difference in incident side: Path difference in reflected side:

$$\Delta X_i = d \sin\theta$$

$$\Delta X_r = d \sin\theta$$

Total path difference; $\Delta X = \Delta X_i + \Delta X_r$
$$n\lambda = d \sin\theta + d \sin\theta$$
$$= 2d \sin\theta \sim 2d\theta$$

Figure 9.2: Right: Glancing of ray at angle θ to the horizontal plane. **Left:** The relation between electron wave length "λ," interplanar distance "d" and glancing angle "θ".

The electron beam coming from specimen is now magnified by objective, intermediate and projector electromagnetic lens. If diffraction pattern is subject of interest, then intermediate electromagnetic lens is focused on the back focal plane of the objective lens. Transmitted ray is blocked, and diffracted rays are allowed from objective aperture, dark field diffraction pattern is formed on the back focal plane of objective. This pattern can be viewed if it is projected on fluorescent screen. Because most of the rays are blocked (transmitted), background of fluorescent screen is dark whereas diffracted rays from atomic edges are allowed so reciprocal lattice pattern[h] of atoms are observed as glowing dots on fluorescent screen.

Now, if transmitted ray is allowed and diffracted rays are blocked from objective aperture, bright field diffraction pattern is formed on the back focal plane of objective. This pattern can be viewed if it is projected on fluorescent screen. Because most of the rays are allowed (transmitted), background of fluorescent screen is bright whereas diffracted rays from atomic edges are blocked so reciprocal lattice pattern of atoms are observed as dark dots on fluorescent screen. The systematic diagram for dark field diffraction pattern and bright field diffraction pattern is presented in Figure 9.3.

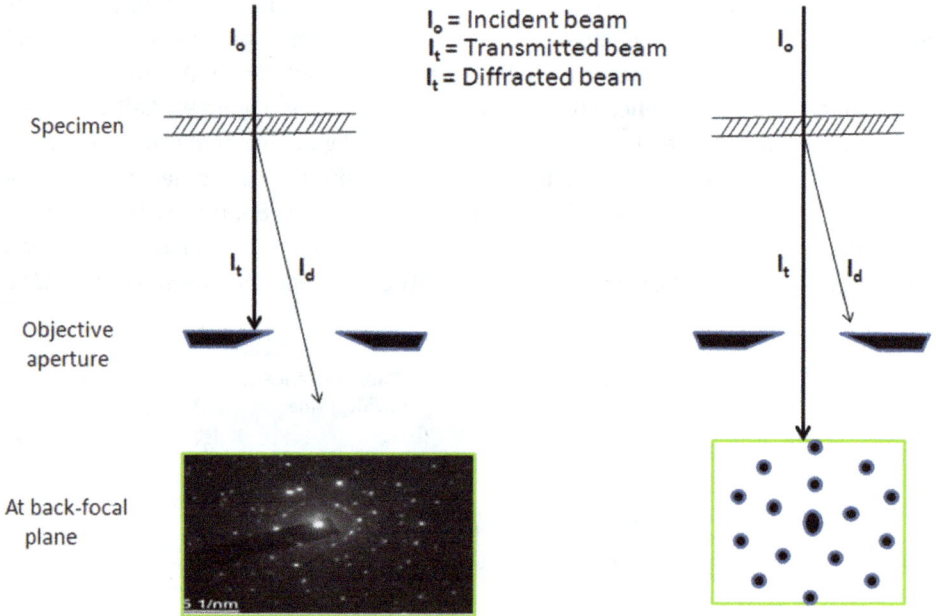

Figure 9.3: Left: Dark field diffraction pattern. Right: Bright field diffraction pattern. With permission from Elsevier.

For image mode, intermediate lens is focused on the image plane of objective electromagnetic lens. Then, all three lens ensure 10^5–10^6 time magnified image of specimen. When electron beam passes through the specimen at any site, it will be deflected as per the mass density (product of density and thickness) of that region. In other word, heavier mass density region deflects more electrons and transmit less electrons. The deflected rays (more than 0.01 radians) are blocked by objective aperture. Overall, fewer rays reach from these regions to fluorescent screen forming darker image for that sites. It can be said that as per the mass-density distribution of site in specimen, brighter to darker image is formed at fluorescent screen (after the focus of objective electromagnetic lens). This is also known as mass-density contrast.

A crystalline specimen with a periodic lattice generates phase difference between the transmitted and diffracted beams. The high-resolution transmission electron microscope (HRTEM) uses both the transmitted and the diffracted beam to create an interference image having periodic dark and bright pattern on the image. HRTEM has been extensively and successfully used for analyzing crystal structures and lattice imperfections in various kinds of advanced materials on an atomic resolution scale. The pictorial presentation of TEM for diffraction pattern and image pattern is shown in Figure 9.4.

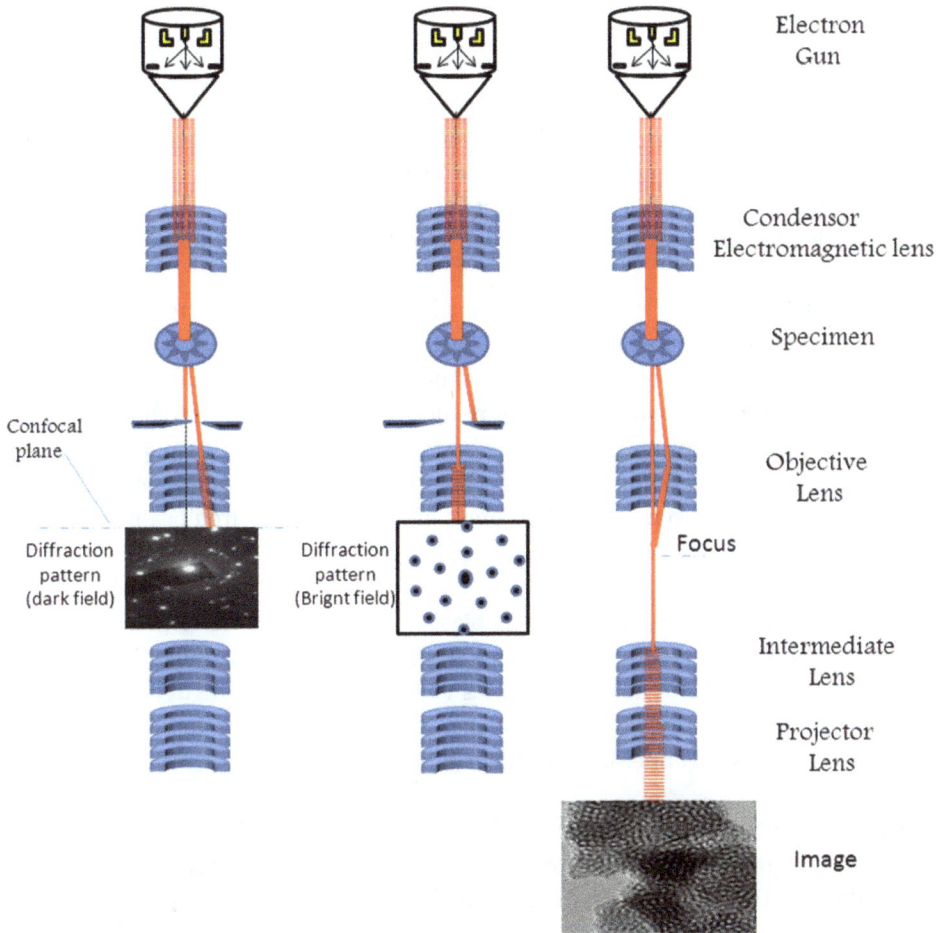

Figure 9.4: The pictorial presentation of TEM for diffraction pattern and image pattern. With permission from Elsevier.

9.3 Glossary

[a]**Resolution:** Resolution is the minimum distance of two points at which they can be visibly distinguished. If a point object is magnified, its image is a central spot surrounded by a series of diffraction rings (airy disk). To see these image points distinctly, airy disks should not severely overlap each other. Resolution (R) of any electron microscope is given by

$$R = \frac{0.61\lambda}{\alpha}; \lambda = \text{wavelength of electron}, \ \alpha = \text{angle of beam deflection}$$

[b]**Magnification:** In TEM, magnification is determined by the power of the objective lens.

[c]**Thermionic emission:** Cathode (made up of tungsten or LaB_6) is heated by electrical current to high temperature (~2,800 K) so that electron gets kinetic energy to overcome the energy barrier and emit out from the cathode surface. The work function of tungsten filament is 4.5 eV whereas it is 2 eV for LaB_6. So, LaB_6 emits high-intensity electron beam and longer life; however, it needs high level of vacuum than tungsten filament.

[d]**Field emission:** Under the influence of very high electric field, electrons are induced to move. If electrons have lack of sufficient energy to cross energy barrier, most of them reflect back from energy barrier and some of them (which has wavelength in the order of the electric field wavelength) bury themselves inside the barrier or cross the barrier or tunnel to the other side of the barrier. Overall, under the applied voltage field, electrons are physically drawn off from a very sharp tip of tungsten (~100 nm). The field emission can generate 10,000 times high-intensity electron beam than tungsten thermionic emission and 100 times than LaB_6 thermionic emission.

[e]**The Wehnelt electrode:** It is negatively biased (–200–300 V). The negatively biased voltage at the Wehnelt electrode creates a small repulsive electrostatic field and the positively biased voltage at anode creates (+1–30 kV) large attractive electrostatic field at the tip of cathode. Overall, under net electric field (due to both attracting anode and repulsive Wehnelt), electron beam of control cross section is pulled out.

[f]**Electromagnetic lens:** It is a coiling of conducting wire which generates magnetic field as per the magnitude of current passes through coil. Electron beam when entering through the region of electromagnetic lens may be deflected by the field magnetic line of force.

[g]**Specimen:** In TEM, specimen preparation is an essential step because of the need of very less specimen thickness (~100 nm) so that specimen can become electronically transparent. By physical cutting (with diamond saw) and hand-grinding, the specimen thickness is reduced to 0.1 mm. Now, specimen (which is good conductor) is put into electrochemical cell (as anode) where acid solution (as electrolyte) is used reduce the specimen thickness up to 100 nm. For metal and ceramic specimen, a beam of energetic ion (energy 1–10 keV) to bombard specimen surface (or ion milling) is also used to reduce the thickness. For polymeric and biological specimen, ultramicrotomy can be used to trim the specimen by diamond knife having sharp cutting tip (1 mm^2 cross section).

[h]**Reciprocal lattice pattern:** In a diffraction pattern, the central spot is for transmitted rays whereas peripheral spots are for diffracted rays. The distance between the central spot of transmitted beam to diffracted spot is not real distance but reciprocal distance. More focus is needed on this subject.

As per Bragg's equation for first order, $\lambda = 2d \sin\theta$; where $\sin\theta = \lambda/2d$. In another way, it can be written as $\sin\theta = (1/d)/(2/\lambda)$. Now, $\sin\theta$ is ratio of perpendicular distances $(1/d)$ and hypotenuse distances $(2/\lambda)$ of right-angle triangle having one acute angle θ. It is right-angle triangle, so these parameters can be easily drawn as circle/sphere elements as hypotenuse distances $(2/\lambda)$ as diameter, $1/d$ as chord of circle and θ is the angle between third side and diameter of the circle/sphere (Figure 9.5). If one ray OP is traced additionally on this geometry and crystal is consider to be located at origin of circle, a very beautiful correlation of incident ray (along AO), transmitted ray (along OQ) and diffracted ray (along OP) at 2θ angle can be observed. The radius of Edward sphere is $1/\lambda$ and the distance QP or $1/d$ is called reciprocal space (d is real space).

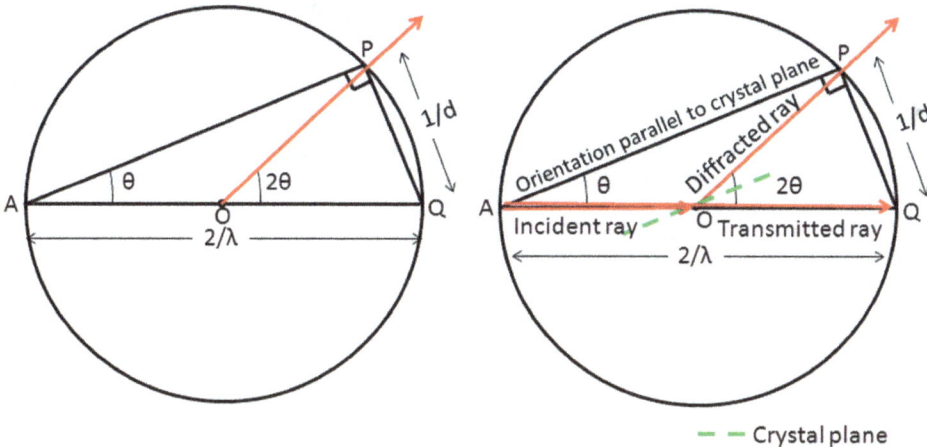

Figure 9.5: A systematic diagram of Edward sphere: Diffraction rays is in line with radius of circle, "incident ray–transmitted ray" is in line with hypotenuse $2/\lambda$ of right-angle triangle (as well as diameter of circle), 2θ is angle between the diffracted and incident ray and perpendicular distance $(1/d)$ of right-angle triangle is chord of circle.

Three dimensionally, if Q is at reciprocal lattice point (000) and P is at reciprocal lattice point (111) then the reciprocal distance is shown by $1/d_{111}$, diffraction angle is θ_{111} and crystal plane is (111). We can say that crystal plane (111) having reciprocal distances lattice $(1/d_{111})$ diffracts the electron beam at angle θ_{111}. Same type of discussion can be carried out for other crystal plane, their reciprocal distances and diffraction brag angle. If all reciprocal distances $(1/d_{hkl})$ concerned with a particular lattice are drawn, then the reciprocal distance pattern of lattice can be calculated.

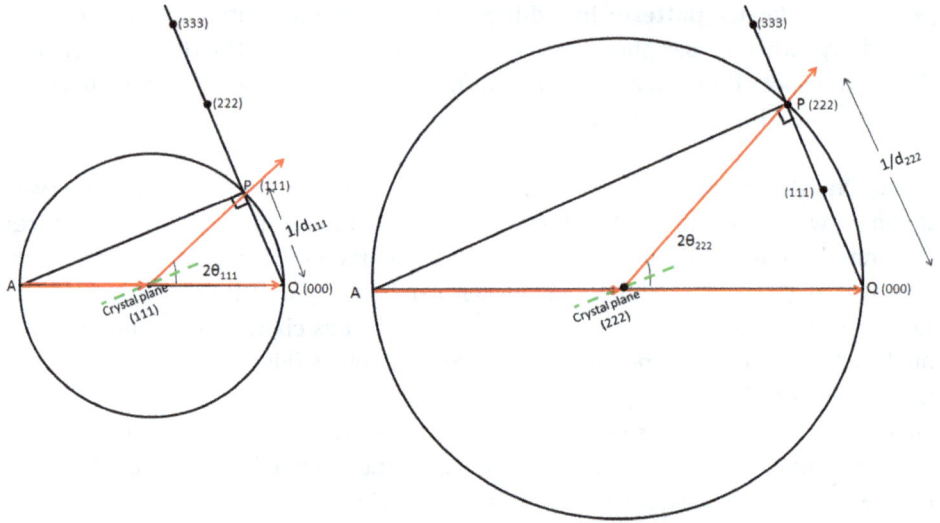

Figure 9.6: Reciprocal distance between two reciprocal lattice points at the Edward sphere.

Reciprocal distance $1/d_{hkl}$ is projected at distance L (camera length) on the fluorescence screen. The diffraction patterns form at fluorescence screen (Figure 9.7). The distance between the central spot of transmitted beam to diffracted spot over fluorescence screen is represented as R (which is definitely reciprocal distance).

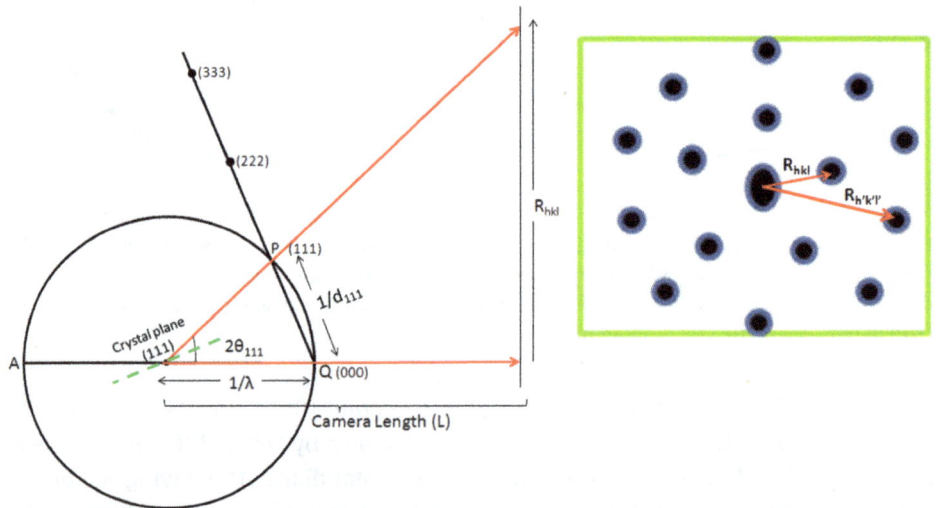

Figure 9.7: Projection of reciprocal distance $1/d_{hkl}$ at cameral length L on the fluorescence screen as central and peripheral dots pattern.

By the geometry of L and 2θ and R can be correlated as

$$L2\theta = R; 2\theta = R/L$$

Putting the 2θ value in TEM Bragg's equation $\lambda = 2d\theta$

$$\lambda L = Rd$$

The reverse information is important for us, because in the fluorescence screen diffraction pattern in TEM, one can easily calculate the distance between the central spot of transmitted beam to diffracted spot. So, ratio of this distant and camera constant (λL) can give reciprocal distance ($1/d_{hkl} = R_{hkl}/\lambda L$). By inverting the reciprocal distance, one can easily know the real distances between crystal planes.

9.4 Analysis of TEM

Example 1: Rahman et al. [1] had prepared Co-incorporated hexagonally ordered mesoporous silica by dissolving poly(ethylene oxide)–poly(propylene oxide)–poly(ethylene oxide) template (P123) dissolved in HCl, followed by n-butanol addition, vigorous stirring with tetraethyl orthosilicates (silica precursor) and cobalt nitrate (cobalt precursor) for 24 h, washing, aged at 100 °C for 24 h in hydrothermal condition, template removal by calcination at 540 °C for 24 h [1]. The sample was abbreviated as Co–HMS–X. The HRTEM image of HRTEM images of 5 mol% Co–HMS–X catalyst is shown in Figure 9.8. The HRTEM images of Co–HMS–X (Co/Si = 5/100) exhibits highly ordered hexagonal array of uniform channels.

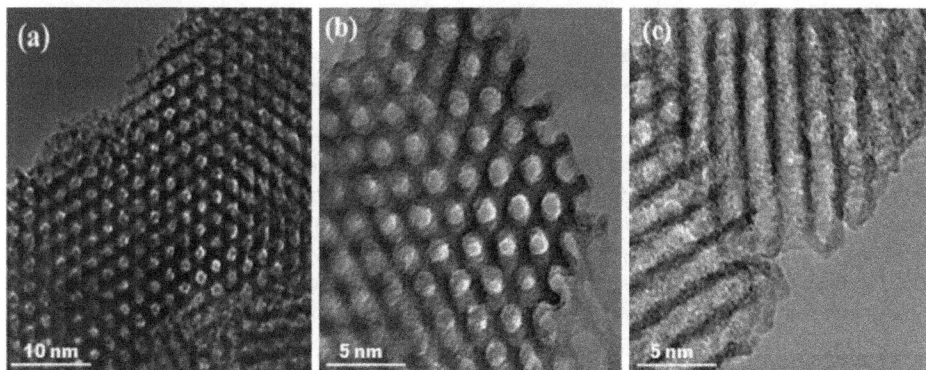

Figure 9.8: The HRTEM image of HRTEM images of 5 mol% Co–HMS–X catalyst. With permission from Elsevier.

Example 2: Patel et al. had prepared 2.5 wt% cerium promoted tungsten-zirconia supported 5 wt% nickel by wet impregnation method using Nickel nitrate hexahydrate and Cerium (III) nitrate hydrate over 9 wt%WO$_3$-91 wt%ZrO$_2$ [2]. The material is abbreviated as 5Ni2.5Ce/WZr. TEM image indicated that the surface topology of the material 5Ni2.5Ce/WZr was quite uniform and ordered. It had ordered channels separated by 0.282-0.330 nm distance.

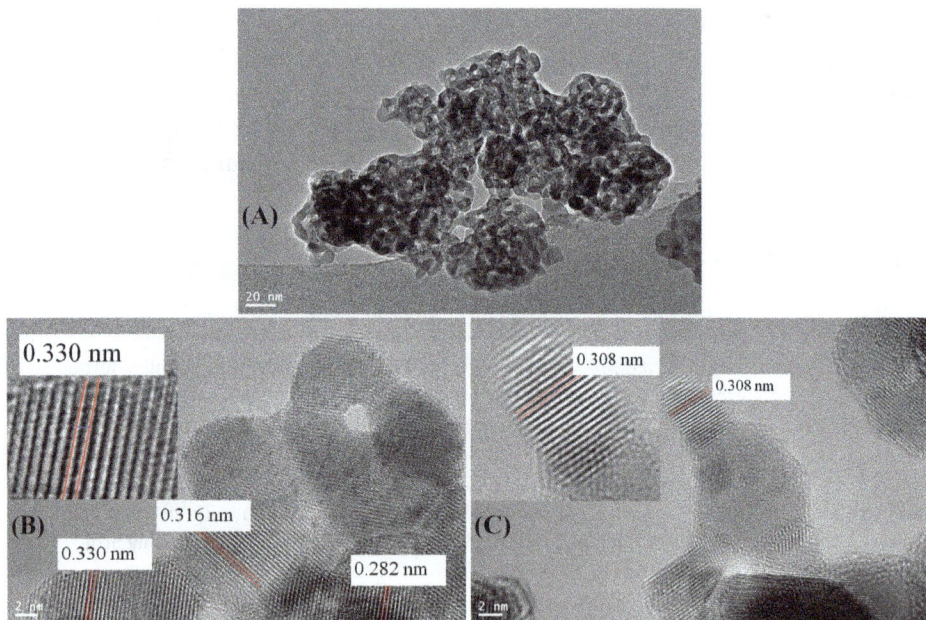

Figure 9.9: TEM micrograph of fresh 5Ni2·5Ce/WZr catalyst. With permission from Elsevier.

Example 3: Mandal et al. [3] had prepared 2.33 wt% gold-impregnated samarium-doped ceria by same above procedure using samarium nitrate hexahydrate as samarium source. The final gel composition for ceria–tin mixed oxide was as follows triethanolamine: Ce(NO$_3$)$_3$ · 6H$_2$O:H$_2$O:Sm(NO$_3$)$_3$ · 6H$_2$O: tetraethylammonium hydroxide = 0.2:0.1:1.1:0.004:0.1 [3]. The sample is leveled as Au/Sm–CeO$_2$ (Sm/Ce = 4/100). The HRTEM image of Au/Sm–CeO$_2$ (Sm/Ce = 4/100) is shown in Figure 9.10. At higher magnification, crystalline particle with a particle size of 30–40 and 8–12 nm was observed in HRTEM. About 10 nm diameter gold nanoparticles having large contact area with the support were easily observable. Lattice fringes were distinctly evident.

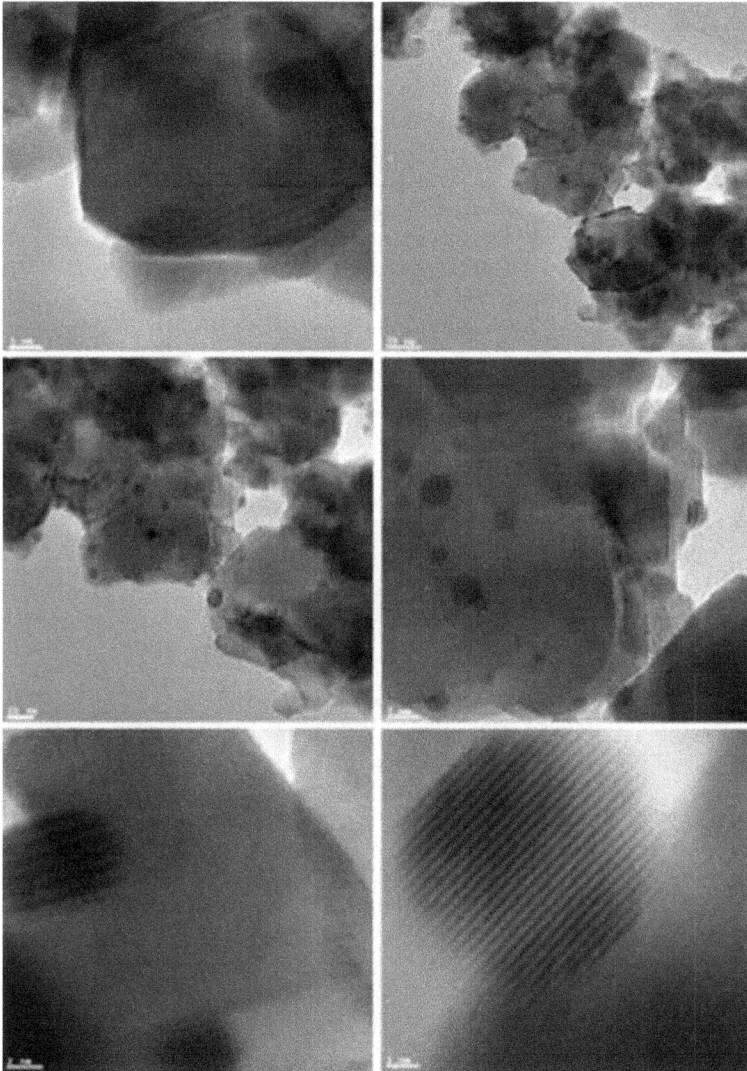

Figure 9.10: High-resolution image of TEM image of Au/Sm–CeO$_2$ (Sm/Ce = 4/100). With permission from Elsevier.

Example 4: Al-Fatesh et al. prepared 5 wt% Ni dispersed over "x wt% 20 lanthana–(100–x) wt% alumina (x = 0, 20, 15, 20)" by mechanical mixing of nickel nitrate, lanthana nitrate and meso-yalumina followed by calcination [4]. The material was utilized for catalyst for the dry reforming of methane. TEM image of fresh and spent sample was taken and shown in **Figure 9.11** The material is abbreviated as 5NixLa(100-x)Al (x = 0, 10, 15, 20 wt%). TEM images showed well dispersion of Ni species over La_2O_3- Al_2O_3 support. Especially over 5Ni15La85Al sample, the average size of Ni species was found around 4 nm (**Figu re 9.11A**) which was increased to 7-8 nm after the reaction (**Figure 9.11B, Figure 9.11C**). 5Ni20La80Al catalyst had also 6.2 nm average size Ni species (Figure of an average size of 6.2 nm (**Figure 9.11A**). Over spent catalyst system 5Ni15La80Al and 5Ni20La85Al, the carbon nanotubes were observed easily (**Figure 9.11C, Figure 9.11D, Figure 9.11F, Figure 9.11G**).

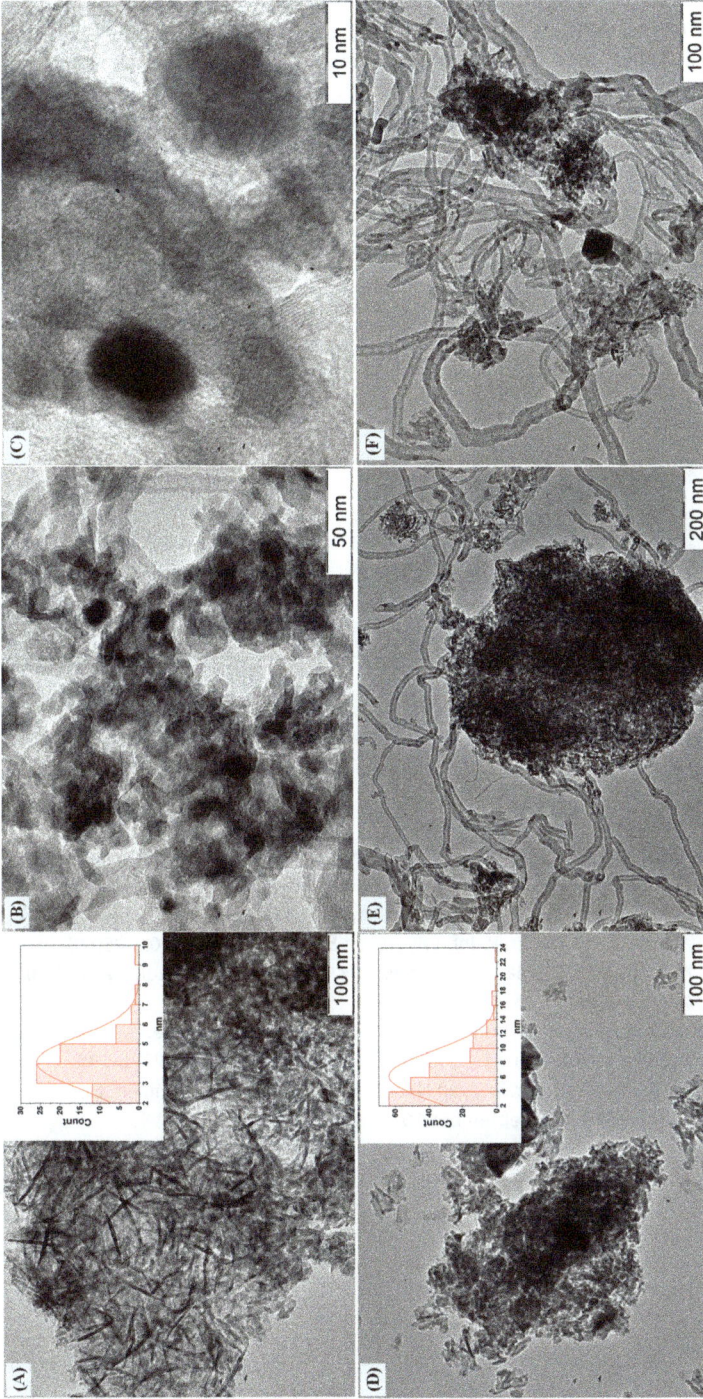

Figure 9.11: TEM image of fresh 5Ni15La85Al catalyst (A) In 100 nm scale (B) 50 nm scale (C) 10 nm scale. (D) TEM image of fresh 5Ni20La85Al catalyst (E) TEM image of spent 5Ni20La85Al catalyst in 200 nm scale (F) TEM image of spent 5Ni20La85Al catalyst in 200 nm scale 100 nm scale. Particle size distribution graph is shown in inset of (A) and (D). With permission from Elsevier.

Example 5: Hassan et al. [5] had prepared cobalt-doped ceria by same above procedure using cobalt nitrate hexahydrate as cobalt source. The final gel composition for cobalt-doped ceria was as follows: TEA:Ce(NO$_3$)$_3$·6H$_2$O:H$_2$O:Co(NO$_3$)$_3$·6H$_2$O:TEAOH = 0.2:0.1:1.1:(0.004−0.008−0.012):0.1 [5]. The sample is leveled as Co(4 mol%)−CeO$_2$. The diffraction pattern of the catalyst confirmed high-level crystallinity (Figure 9.12).

5 1/nm

Figure 9.12: The diffraction pattern of Co(4 mol%)−CeO$_2$ catalyst. With permission from Elsevier.

References

[1] Rahman, S., Santra, C., Kumar, R., Bahadur, J., Sultana, A., Schweins, R., Sen, D., Maity, S., Mazumdar, S., Chowdhury, B. *Appl. Catal. A Gen.* **2014**, *482*, 61–68.
(2) Patel, R., Al-Fatesh, A. S., Fakeeha, A. H., Arafat, Y., Kasim, S. O., Ibrahim, A. A., Al-Zahrani, S. A., Abasaeed, A. E., Srivastava, V. K., Kumar, R. Int. J Hydrogen Energy 2021, 46, 25015-25028
[3] Mandal, S., Bando, K. K., Santra, C., Maity, S., James, O. O., Mehta, D., Chowdhury, B. *Appl. Catal. A Gen.* **2013**, *452*, 94–104.
(4) Al-mubaddel, F. S., Kumar, R., Lanre, M., Frusteri, F., Aidid, A., Kumar, V., Olajide, S., Hamza, A., Elhag, A., Osman, A. I., Al-Fatesh, A. S. Int. J. Hydrogen Energy 2021, 46(27),14225–14235
[5] Hassan, S., Kumar, R., Tiwari, A., Song, W., van Haandel, L., Pandey, J. K., Hensen, E., Chowdhury, B. *Mol. Catal.* **2018**, *451*(September 2017), 238–246.

10 Scanning tunneling microscopy

10.1 Background

Where TEM and SEM techniques are based on far-field interaction between electrons and specimen, scanning tunneling microscopy (STM) is based on near-field interaction between the detector tip and specimen. In STM, the detector tip is away from the specimen only by atomic distances. In this close proximity, there is no possibility of formation of airy disk (due to electron diffraction rings) between two points. So, the resolution limit due to diffraction rings (between two distinct points) is eliminated, and the true image of surface atoms has been formed with the lateral range up to about 100 μm and vertical range up to about 10 μm. STM is operated in vibration-free environment because the detector tip (probe) is only atomic distance away from the examined surface. To eliminate floor vibration, SPM is mounted on spring, rubber feet or air-damped feet, whereas for acoustic vibration the isolation instrument is operated in a solid box.

10.2 Instrumentation and working principle

The systematic diagram of STM is shown in Figure 10.1. On a surface, the wave function $\mathbf{\Psi}$ can be shown as follows:

$$\mathbf{\Psi}(x) = \exp\left(-\beta x\right) \text{ where } \beta = \frac{\sqrt{2m}\,\Phi}{\hbar} \text{ and } \Phi \text{ is work} - \text{function.}$$

There will be no electron current out of the surface, or the wave function outside the metal will be classically forbidden. If one brings up a metal probe within a few atomic dimensions of the surface, one will form a quantum barrier between the two conducting regions. Or it is through regions halted by some small barriers. Then it is possible for electrons to a quantum tunnel from the sample to the probe.

So, under the applied voltage field between specimen (under examination) and tungsten probe, electrons are physically drawn off from specimen to probe. The amount of current which buried themselves inside the quantum barrier or cross the quantum barrier or tunnel to the other side of the quantum barrier is known as the tunneling current as per the following equation:

$$I_t = V_b \exp^{(-Cd)}$$

where I_t is the tunneling current, V_b is the applied voltage, C is constant that depends on the nature of the sample material and d is the distance between probe tip and sample.

https://doi.org/10.1515/9783110656480-010

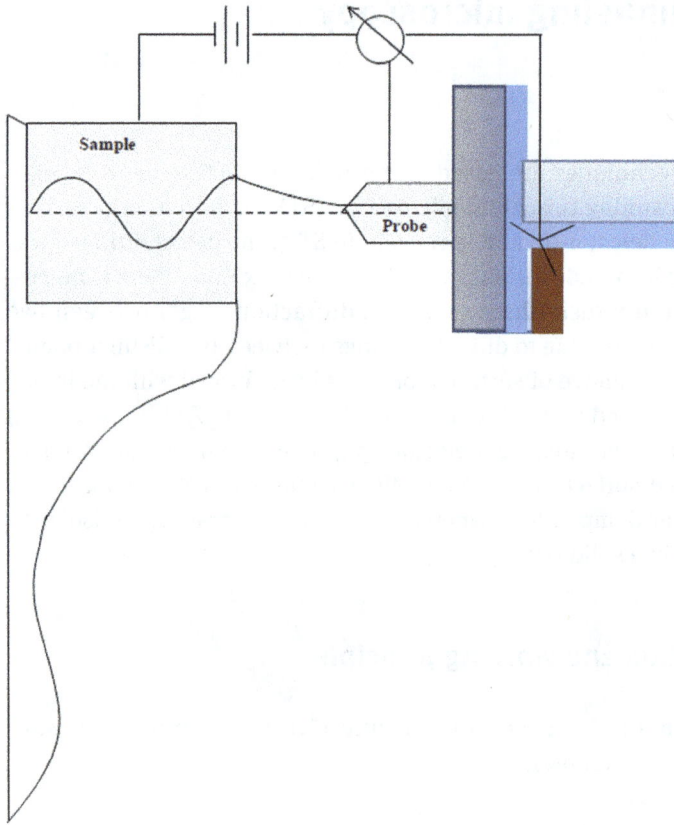

Figure 10.1: The systematic diagram of scanning tunneling microscope.

So, the tunneling current purely depends on the bias voltage applied and the distance between the surface and probe tip. As a rough approximation to generate a tunneling current (between 10 pA and 10 nA), only small bias voltage (1 mV–4 V) is needed if the gap is on the scale of interatomic distance.

To maintain the constant tunneling current during the scanning, probe tip is mounted on three piezoelectric tubes[a] (arrange in an orthogonal arrangement). Further, the tunneling current is used as the input signal for feedback loop. The output signal of feedback loop is amplified to a necessary electric potential to activate the piezoelectric scanner.

Overall, to maintain the constant current, scanner is moving up and down. This means a corrugation contour of local density of states (LDOS)[b] that is sensitive to the atom location at the Fermi level[c] is obtained in constant current mode.

If feedback loop is turned off, the scanner will not move up and down as per the input tunneling current. This mode is called constant height mode. It scans in much higher rates so it is useful for observing time-dependent dynamic processes.

But due to high speed, there is a risk of crashing the tip onto the sample surface during scanning. If tip is stopped at a position above the sample and tunneling current is recorded by changing either the height of tip and sample or voltage, then this mode is a spectroscopic mode. It is useful in understanding superconduction and molecular adsorption on metal.

The adatom may be interacted weakly by a probe tip. So, under the weak interaction with probe tip, adatoms can move laterally on the specimen surface. It is called lateral manipulation. Again if adatom may interact strongly toward the tip, adatoms may detach from the surface and attach to the tip under the voltage pulses. Further, it may be deposited at another location on the surface (detachment from the tip to the surface) under the voltage pulses. This is called vertical manipulation.

10.3 Glossary

[a]**Piezoelectric tube:** The piezoelectric material changes its relative shape as per the applied voltage as follows:

$$\frac{\Delta L}{L} = d_{ij}E$$

E is the applied potential, d_{ij} is the piezoelectric coefficient and its typical value is about 0.3 nm mm/V:

$$\frac{\Delta L}{EL} = d_{ij}$$

$$\frac{1\mu m}{300 \times 10 \, mm} = 0.3$$

Roughly if d_{ij} is 0.3 nm mm/V, 300 V potential is required to achieve 1 μm movement of a 10 mm long piezoelectric tube. For three-dimensional positioning of scanner, such three piezoelectric tubes in an orthogonal arrangement is needed. In this way, the tunneling current always gives input to precisely control the motion of a piezoelectric scanner.

[b]**LDOS:** The LDOS is the number of electronic states per unit energy range over a surface. As for example, Besenbacher et al. showed that when nickel(111) surface is substituted by gold, (1) LDOS is lower at the substituted gold atoms position than on the adjacent nickel. Due to local modification of the electronic structure of Ni, the surrounding Ni sites appeared brighter and Au atoms appear black in STM image. Over et al. showed that (2) the array of bridging oxygens along the [001] direction over $RuO_2(110)$ surface was viewed as bright strings in STM image. On exposure of CO over $RuO_2(110)$ surface, oxygen is consumed, oxygen vacancies are formed and

CO is oxidized to CO_2. In the STM image, loss of bridging oxygen was seen as loss of bright strings density and dark strips are viewed as defects. Due to the loss of bridging/linking O from Ru, Ru atoms form clusters and these clusters are viewed as bright protrusions in an STM image.

[c]**Fermi level:** The Fermi level is the highest energy level occupied by an electron in the metal at absolute zero temperature.

[b,c]**LDOS at Fermi level:** LDOS at Fermi-level information is important to know the numbers of electrons that can participate in electron conduction. An adsorbed molecule over the surface may reduce LDOS.

10.4 Analysis of STM image

Example 1: Hu et al. made 2 nm well-ordered $CeO_2(111)$ thin film grown on Cu(111). Further, different amounts of metallic Zr were deposited over $CeO_2(111)$ with up to 0.5 ML at 300 K [3]. At 0.4 ML coverage, small Zr clusters (average height: 0.14 ± 0.08 nm; average diameter: 1.08 ± 0.03 nm; and cluster density: $8.1 \pm 0.3 \times 10^{12}/cm^2$) were observed in STM image (Figure 10.2a). Up to 0.06–0.08 ML coverage, the cluster size remained the same (average height: 0.14 ± 0.05 nm; average diameter: 1.30 ± 0.40 nm) and the number of clusters increased (Figure 10.2b, c). The aspect ratio (height/diameter) remains about 0.11 up to 0.12 ML coverage (Figure 10.2d). It indicated that Zr had strong interaction with $CeO_2(111)$ surface; it did not migrate on $CeO_2(111)$ surface to form relatively large Zr clusters up to 0.12 ML coverage. On 0.5 ML coverage, the cluster size (average height: 2.01 ± 0.43 nm; average diameter: 0.19 ± 0.05 nm; and cluster density: $25.1 \pm 0.7 \times 10^{12}/cm^2$) had grown eventually, and about 70% of the CeO_2 surface was covered (Figure 10.2e, f).

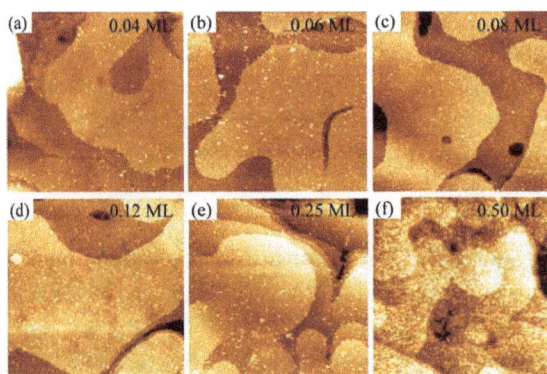

Figure 10.2: STM images of Zr deposited on 2 nm CeO_2 films with different coverages. With permission from American Chemical Society.

Example 2: Choi et al. prepared 0.35 nm ZrO_2 ultrathin films by oxidation of $Pt_3Zr(0001)$ and Pd_3Zr (0001) followed by annealing at 950 °C. Further, Ag was deposited over ZrO_2 (coverage 0.26 ML) by the evaporation method [4]. At the steps (marked under cyan dashed–dotted lines) and domain boundaries (marked under broken yellow lines), high density of Ag clusters (apparent height: 0. 8–2.5 nm) was observed in STM image in Figure 10.3a. However, Ag did not nucleate if the oxide lattice was only slightly distorted there (shown by white arrow in Figure 10.3b).

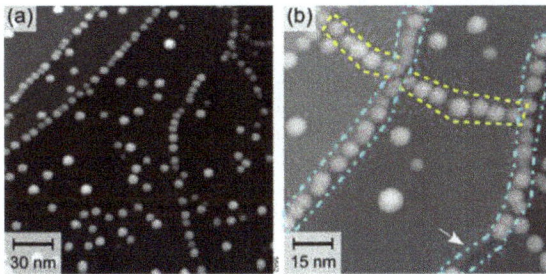

Figure 10.3: (a, b) STM images of Ag on ZrO_2 films over Pt_3Zr (ZrO_2/Pt_3Zr) with 0.26 ML coverage at room temperature. With permission from American Chemical Society.

Example 3: J. Kibsgaard et al. prepared MoS_2 nanoparticle by evaporating Mo metal over single-crystal rutile $TiO_2(110)$ followed by high-temperature sulfidation step [5]. In STM image of TiO_2 showed Ti troughs as the bright rows, bridging O rows were imaged as dark rows and bridging hydroxyl groups were imaged by bright protrusions (Figure 10.4A). MoS_2 has semiconducting property. So, the apparent height of MoS_2 in STM image depends on applied bias voltage. On 1,250 mV bias voltage, the height of the hexagonal basal plane of the MoS_2 nanoparticles is 4.7 ± 0.5 Å relative to the $TiO_2(110)$ surface. The dimension of a single S–Mo–S layer (core-to-core distance) is 3.15 Å (Figure 10.4B), which means single-layer MoS_2 nanoparticles were formed on the TiO_2 surface. At low-bias voltages, STM primarily tunnels to electronic states inside the band gap of MoS_2 and perturbation to charge transfer from the bright Ti rows of the TiO_2 substrate to the MoS_2 particle happens. So, at 883.5 and 46.4 mV bias voltage, STM showed the characteristic row structure of the $TiO_2(110)$ surface on the basal plane of the MoS_2 nanoparticles (Figure 10.4C, D)

Example 4: Büchner et al. had prepared a bilayer silica film on Ru(0001) single crystal (with 1.6 ML coverage) through physical vapor deposition and subsequent annealing under ultrahigh vacuum [6]. In its earlier publication, Büchner showed that lognormal ring size distribution (occurrence of different ring sizes) is a general feature of amorphous networks [7]. In STEM image, such amorphous network pattern of silica film was observed (Figure 10.5a, b). Conventionally, 1 ML and 2 ML coverage referred to a monolayer and bilayer silica film, respectively. So here, 1.6 ML coverage showed flat sheet of two layers of thickness exhibiting holes (up to 30 nm in diameter) (Figure 10.5c, d).

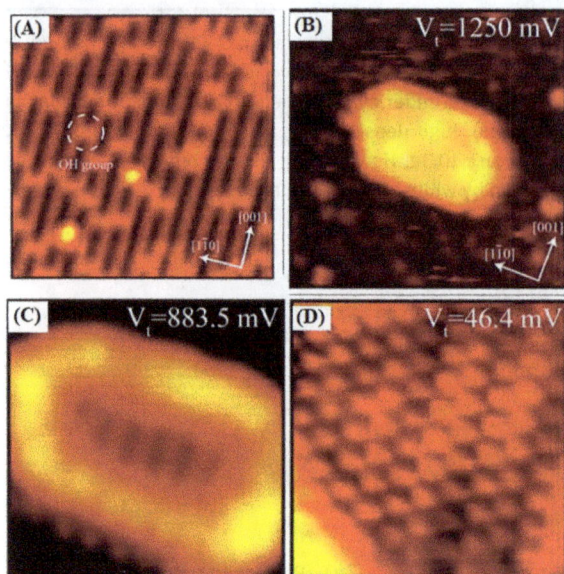

Figure 10.4: STM image of (A) TiO_2(110) surface, (B) MoS_2 nanoparticles on a TiO_2(110) surface at 1,250 mV, (C) MoS_2 nanoparticles on a TiO_2(110) surface at 883.5 mV and (D) MoS_2 nanoparticles on a TiO_2(110) surface at 46.4 mV. With permission from Elsevier.

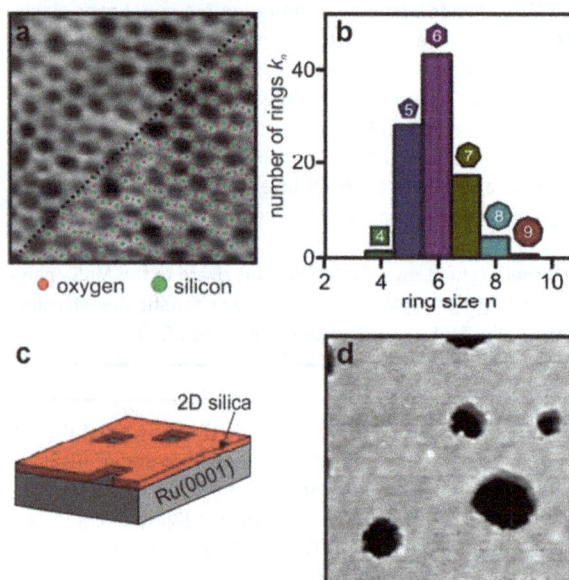

Figure 10.5: (a) Atomic-resolution STM image of an amorphous silica bilayer on Ru(0001). (b) Ring size histogram of the amorphous network shown in panel. (c) Schematic picture of partial coverage of the silica bilayer. (d) Large-scale STM image showing partial coverage of the silica bilayer (overall coverage 1.8 ML). With permission from American Chemical Society.

Example 5: Berner et al. prepared ultrathin cerium films over Pt(111) by evaporation of metallic cerium onto clean Pt(111) [8]. The first step is cleaning of platinum single crystal by repeated sputtering with argon for 20 min followed by heating in oxygen for 15 min and annealing at 1,100 K for 1 h in ultra-high vacuum. The height of clean single step was found 0.64 ± 0.1 Å [9]. The second step is evaporation of metallic cerium onto clean Pt(111). STM images showed that Ce islands (40–80 Å diameter and 15 Å height) were scattered over Pt(111) surface with an apparent preference (Figure 10.6 left). Now on subsequent heating of sample at 900 K, 20 ceria aggregated islands were at and over step edges. The step height of ceria island was 2.26 Å more than clean Pt single steps (Figure 10.6 right).

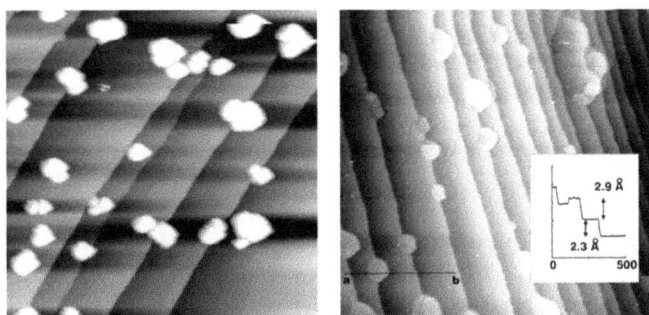

Figure 10.6: Left: Constant-current STM image of Ce dosed Pt(111) surface. Right: Constant-current STM image of Ce dosed Pt(111) surface after heating at 900 K under ultra-high vacuum. With permission from Elsevier.

References

[1] Besenbacher, F., Chorkendorff, I., Clausen, B. S., Hammer, B., Molenbroek, A. M., Nørskov, J. K., Stensgaard, I. *Science (80-.)*. **1998**, *279*(5358), 1913–1915.

[2] Over, H., Kim, Y. D., Seitsonen, A. P., Wendt, S., Lundgren, E., Schmid, M., Varga, P., Morgante, A., Ertl, G. *Science (80-.)*. **2000**, *287*(5457), 1474–1476.

[3] Hu, S., Wang, W., Wang, Y., Xu, Q., Zhu, J. *J. Phys. Chem. C* **2015**, *119*(32), 18257–18266.

[4] Choi, J. I. J., Mayr-Schmölzer, W., Ilaria Valenti, P. L., Mittendorfer, F., Redinger, J., Diebold, U., Mael S. **2016**, 9920–9932.

[5] Kibsgaard, J., Clausen, B. S., Topsøe, H., Lægsgaard, E., Lauritsen, J. V., Besenbacher, F. *J. Catal.* **2009**, *263*(1), 98–103.

[6] Büchner, C., Wang, Z. J., Burson, K. M., Willinger, M. G., Heyde, M., Schlögl, R., Freund, H. J. *ACS Nano* **2016**, *10*(8), 7982–7989.

[7] Büchner, C., Schlexer, P., Lichtenstein, L., Stuckenholz, S., Heyde, M., Freund, H. J. *Zeitschrift Fur Phys. Chemie* **2014**, *228*(4–5), 587–607.

[8] Berner, U., Schierbaum, K. *Thin Solid Films* **2001**, *400*(1–2), 46–49.

[9] Search, H., Journals, C., Contact, A., Iopscience, M., Address, I. P. *207*, 5–6.

Annexure I

d^1 system (comparable d^1–d^9 and d^6–d^4 system)

The ground state configuration of d^1 and d^9 can be written as

d^1: [↑ | | | |]
m_i=+2, +1, 0, -1, -2

d^9: [↑↓ | ↑↓ | ↑↓ | ↑↓ | ↑]
m_i= +2, +1, 0, -1, -2

Electrons are spinning about their axis with spin momentum $s_1 = \frac{1}{2}$. Total spin of d^1 or d^9 will be $\Sigma s = \frac{1}{2}$. In magnetic field, every state orients in $2S + 1$ manner along magnetic field. It is called multiplicity. So, multiplicity is "M" $= 2S + 1 = 2(1/2) + 1 = 2$. Again, electrons are revolving around nucleus in d orbital where an electron has a total sum of angular momentum 2; $\Sigma m_l = l = 2$. Different angular momentum states are represented by capital letter of alphabet G, F, D, P and S in place of numerical value 4, 3, 2, 1 and 0, respectively. (Note: In the same manner, small alphabets g, f, d, p and s are used for 4, 3, 2, 1 and 0 orbital quantum number respectively.) So, this state is shown by D. Overall total momentum states (energy state) can be shown by angular momentum states (L) and multiplicity (M) as $^M L$. For d^1, the total momentum state is $^2 D$.

The ground state configuration of d^6 and d^4 can be written as

d^6: [↑↓ | ↑ | ↑ | ↑ | ↑]
m_i= +2, +1, 0, -1, -2

d^4: [↑ | ↑ | ↑ | ↑ |]
m_i= +2, +1, 0, -1, -2

Electrons are spinning about their axis with spin momentum $s_1 = \frac{1}{2}$. Total spin of d^6 or d^4 will be $\Sigma s = (\frac{1}{2}) \times 4 = 2$. In magnetic field, every state orients in $2S + 1$ manner along magnetic field. It is called multiplicity. So, multiplicity is $M = 2S + 1 = 2(2) + 1 = 5$. Electrons are revolving around nucleus in d orbital where an electron has a total sum of angular momentum 2; $\Sigma m_l = l = 2$. Overall, total momentum states (energy states) can be shown by angular momentum states and multiplicity as $^M L$ as $^5 D$.

Splitting of D energy state in the unsymmetrical field (octahedral ligand field)

"d" orbital in octahedral field splits into interaxial orbital lobes t_{2g} (lower) and axial orbital lobes e_g (higher). In ground state, one electron is grazing in interaxial lobes of "d" orbitals. There may be three such possibilities. So, it is triply degenerate (T). If a rotation axis in "z" direction is fitted and rotated up to 180°, then same lobes

https://doi.org/10.1515/9783110656480-011

are generated. So, it has $360°/180° = 2$ or second-order rotation (2). All lobes belong to "d" orbital. So, these are gerade (g).

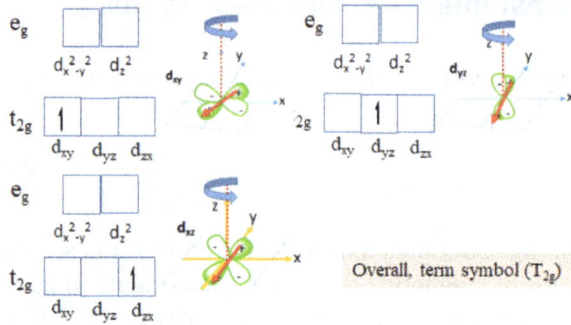

In excited state, one electron is grazing in axial lobes of "d" orbital. There may be two such possibilities. So, it is doubly degenerate (E). As it is related to "d" orbitals, it is also gerade (g).

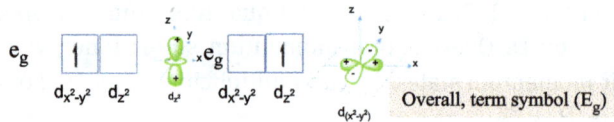

On increasing ligand field strength, gap between split states increases. It is shown by the Orgel diagram.

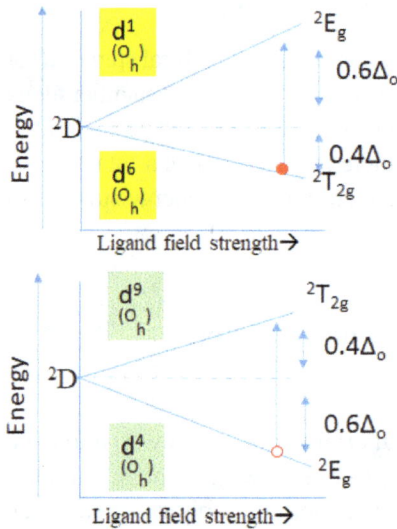

Note: d^n and $d^{(5+n)}$ (where $n < 5$) electron has similar angular momentum. So, d^1 and d^6 electrons in the octahedral field have equivalent energy state and splitting pattern.

In the case of d^9, one hole (hole means how much deficient from full) is present in E_g energy level (it is upper energy state for electron system but ground state for hole system). On providing energy, hole is transferred from E_g to T_{2g} energy level. So, d^n and d^{10-n} systems have splitting pattern opposite to each other, that is, d^1 vis-à-vis d^9 as well as d^6 vis-à-vis d^4 have opposite splitting pattern in octahedral field. It can be represented by common Orgel diagram.

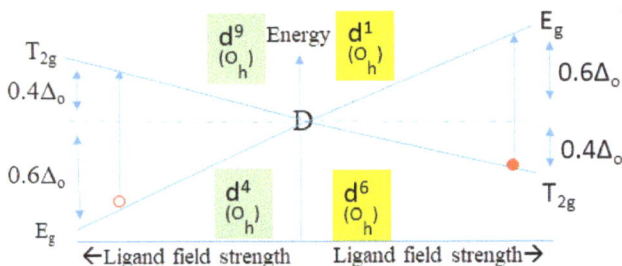

So, on providing energy, there will be single transition of electron from ground state to excited state in the d^1 case. It is observed in absorption spectra of $[Ti(H_2O)_6]^{3+}$ at 20,300 cm^{-1}. As strength of ligand field increases, gap between energy level increases and so absorption frequency increases as $[TiCl_6]^{3-}$ at 13,000 cm^{-1}, $[TiF_6]^{3-}$ at 18,900 cm^{-1}, $[Ti(H_2O)_6]^{3+}$ at $20,300$ cm^{-1} and $[Ti(CN)_6]^{3-}$ at $22,300$ cm^{-1}.

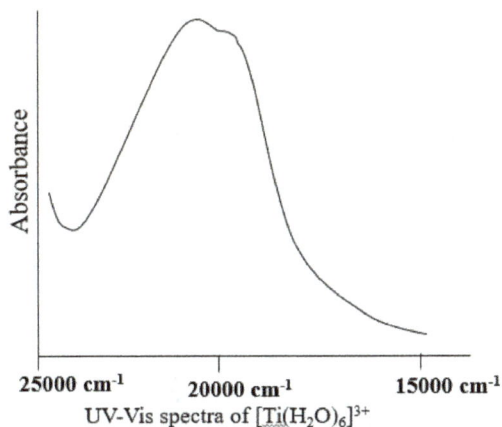

UV-Vis spectra of $[Ti(H_2O)_6]^{3+}$

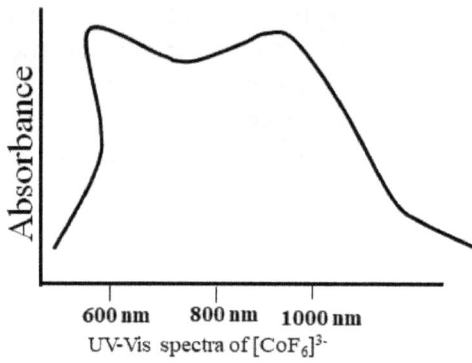

UV-Vis spectra of $[CoF_6]^{3-}$

UV spectra of d^1 complex show a single peak with a shoulder due to Jahn–Teller distortion. Same UV spectra of d^6 complex shows two clear peaks due to Jahn–Teller distortion (discussed in detail in Jahn–Teller section).

Annexure II

d^2 system (comparable d^2–d^8 and d^3–d^7 system)

For more than one electron system, electron–electron "spin-wise (spin momentum)" interaction and "orbit-wise (angular momentum)" interaction defines different energy states as well as different arrangement of electrons in suborbital.

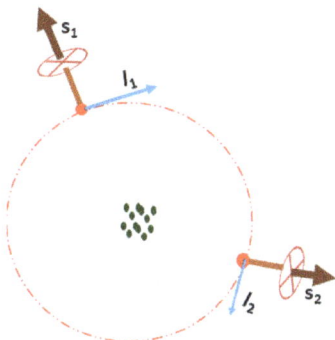

Spin (S): Atom is made up of central nucleus and peripheral electrons. Electrons are spinning about its axis with spin momentum $s_1 = 1/2$, $s_2 = \frac{1}{2}$. The total spin angular momentum has different state (spin states) as $S = \Sigma s$, $(\Sigma s - 1)$ . . . .up to $0 = 1, 0$. In magnetic field, every state orients in $2S + 1$ manner along magnetic field. It is called multiplicity. So, it is either 3 or 1; $M_1 = 2(1) + 1 = 3$, $M_2 = 2(0) + 1 = 1$.

Angular momentum (L): Electrons are revolving around nucleus in "d" orbital. "d" orbital has angular momentum 2. So, every electron has angular momentum 2; $l_1 = 2$, $l_2 = 2$. The resultant angular momentum has different states as Σm_l, $\Sigma m_l - 1$, up to $0 = 4, 3, 2, 1, 0$. Different angular momentum states are represented by capital letter of alphabet G, F, D, P and S in place of numerical value 4, 3, 2, 1 and 0 respectively. (Note: In the same pattern, small alphabets g, f, d, p and s are used for 4, 3, 2, 1 and 0 orbital quantum number respectively.) Overall total energy state can be shown by angular momentum states and multiplicity as $^M L$ as here, ^3G, ^1G, ^3F, ^1F, ^3D, ^1D, ^3P, ^1P, ^3S and ^1S.

As per different angular momentum, electron locates in different suborbital d_{xy}, d_{yz}, d_{zx}, d_{x2-y2}, d_{z2} having momentum $|\ m_l\ | = |\ +2, +1, 0, -1, -2\ | = 2, 1, 0$. As per the Hunds' rule, two electrons of same spin can be placed in different suborbitals, whereas two electrons of opposite spin should be placed in same suborbital.

https://doi.org/10.1515/9783110656480-012

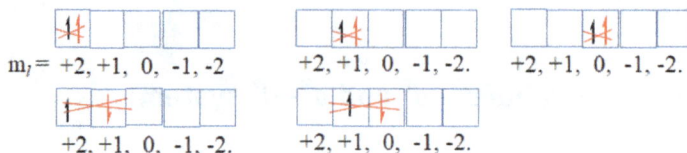

$$m_j = +2, +1, \ 0, \ -1, -2 \qquad +2, +1, \ 0, \ -1, -2. \qquad +2, +1, \ 0, \ -1, -2.$$

$$+2, +1, \ 0, \ -1, -2. \qquad +2, +1, \ 0, \ -1, -2.$$

^3G, ^3D and ^3S are only possible if two electrons of same spin will reside in a single suborbital, which violate the Hunds' rule. ^1F and ^1P are only possible if two electrons of opposite spin will reside in different suborbitals, which also violate the Hunds' rule. Overall, the allowed energy states can be shown as ^1G, ^3F, ^1D, ^3P, ^1S.

Energy states

(a) Stability on the basis of spin states

If two electrons of same spin are placed in different suborbitals, then both electrons are distant and they repulse each with less intensity. So, higher spin state (or triplet state) is more stable. So, among different momentum states (^1G, ^3F, ^1D, ^3P, ^1S), triplet states (^3F, ^3P) are more stable than singlet states (^1G, ^1D, ^1S).

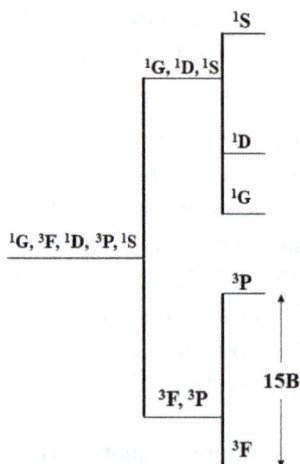

(b) Stability on the basis of angular momentum states

Electrons having higher angular momentum cover large regions in atomic space which ensure least electron–electron repulsion. So, among orbitals with same multiplicity, electron which has higher angular momentum is more stable.

As among ^3F and ^3P, ^3F is more stable than ^3P.

^{3}F: | ↑ | ↑ | | | |

m_l = +2, +1, 0, -1, -2.

^{3}P: | | ↑ | ↑ | | |

m_l = +2, +1, 0, -1, -2.

Among singlet multiplicity (^1G, ^1D, ^1S), ^1G is more stable than ^1D and ^1S. The energy gap (separation) between energy states is shown in the order of Racah parameter "B." Racah parameter is a measure of interelectronic repulsion. It is derived from linear combination of exchange integrals and coulomb integrals of free ion (or uncomplexed ion). The data for Racah parameter is already available in textbook for each type of uncomplexed ion. In the case of Cr^{3+} (d^2) case, the separation between ^3F and ^3P energy states is $15B$.

(c) Organization of unsymmetrical field about ion

Metal ion has electrons localized over itself. When ligands are approaching about metal ion, metal–ligand complex is formed. Then, electrons on metal are delocalized in the region of metal–ligand system. So, interelectronic repulsion among distant electrons decreases. This effect is called Nephelauxtic effect. Nephelauxtic effect also indicates the development of covalent character among metal–ligand system. Due to change in interelectronic repulsion, the value of Racah parameter is again revised (as per new linear combination of exchange integrals and coulomb integrals of complexed ion). It is called apparent Racah parameter B′. Due to unsymmetrical approach of ligands, all energy states component experience unequal force which leads further splitting of energy states. The split states are called energy levels.

(d) Splitting of energy state in the unsymmetrical field

In unsymmetrical field, **P state** is not split and it transforms into T_{1g}. For clarity from similar F state spitting component, it is shown as $T_{1g}(P)$. In unsymmetrical field, **F energy state** splits into various energy levels. "d" orbital in octahedral field splits into interaxial orbital lobes t_{2g} (lower) and axial orbital lobes e_g. If one electron is excited from t_{2g} to excited state e_g, then the most stable arrangement of electrons are those set where electrons are placed far apart or **right angle to each other** so that minimum interelectron repulsion takes place. There may be three such possibilities. So, it is triply degenerate (T).

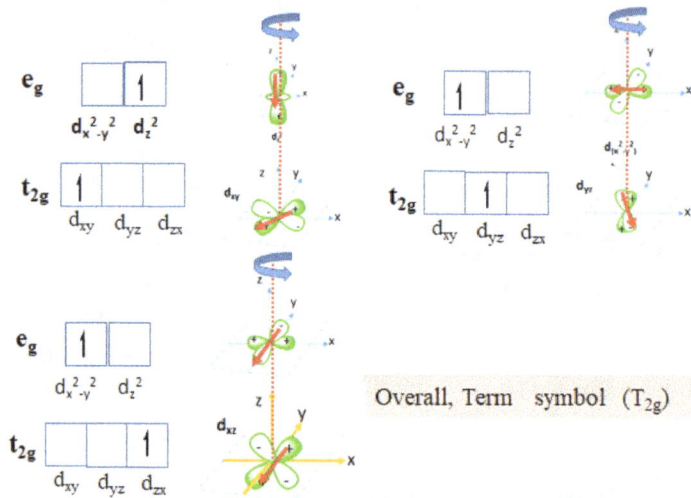

Overall, Term symbol (T_{2g})

If a rotation axis in "z" direction is fitted and rotated up to 180°, then same lobes are generated. So, it has 360°/180° = 2 or second-order rotation (2). All orbitals belong to "d" orbital, so they are gerade (g). If one electron is excited from t_{2g} to excited state e_g, and electrons are at 45° to each other and coplanar, then interelectron repulsion increases. So, this state will be less stable. There may be three such possibilities. So, it is triply degenerate (T). If a rotation axis in "z" direction is fitted and rotated up to 360°, then only same lobes are generated. So, it has 360°/360° = 1 or first-order rotation (1). All orbitals belong to "d" orbital, so they are gerade (g).

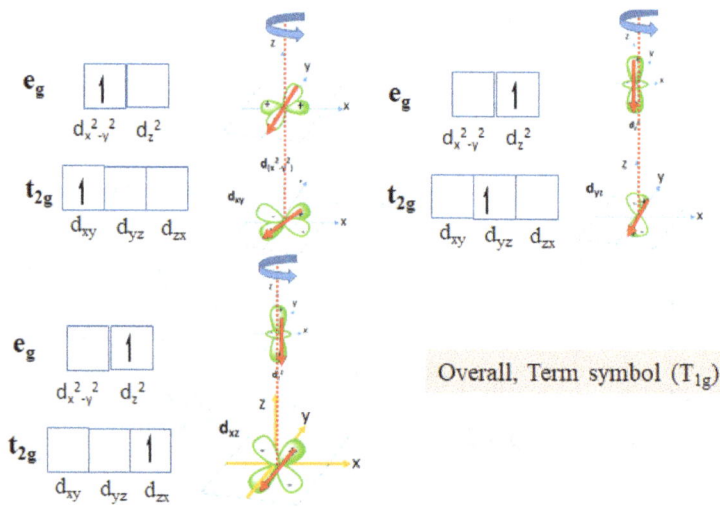

Overall, Term symbol (T_{1g})

If both electrons are excited from t_{2g} to e_g, the energy gap disappears and electrons strongly repel each other. So, it is least stable. It is singly degenerate (A), second-order rotation (2) and gerade (g).

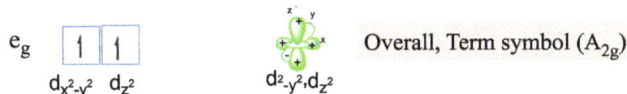

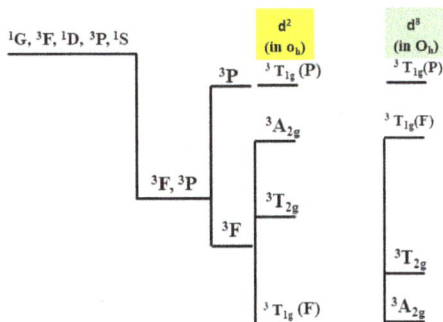

e_g | ↑ | ↑ |

$d_{x^2-y^2}$ d_{z^2}

$d_{z^2-y^2}, d_{z^2}$ Overall, Term symbol (A_{2g})

$^1G, ^3F, ^1D, ^3P, ^1S$		d^2 (in o_h)	d^8 (in O_h)
	^3P	$^3T_{1g}$ (P)	$^3T_{1g}$(P)
		$^3A_{2g}$	$^3T_{1g}$(F)
$^3F, ^3P$		$^3T_{2g}$	
	^3F		$^3T_{2g}$
		$^3T_{1g}$ (F)	$^3A_{2g}$

(e) Spectra

Under the spin selection rule, $\Delta S = 0$ (during transition, electron does not change its spin). So, among all, only transition between ^3F and ^3P are possible. It is spin-allowed transition. In octahedral field, ^3F is split into A_{2g}, T_{2g} and T_{1g}, but ^3P is not split (discussed earlier). P state transforms into T_{1g}. Among energy levels (split level) of ^3F, overall T_{1g} is more stable than T_{2g} and A_{2g}. For clarity between T_{1g} of ^3P and T_{1g} of ^3F, generally T_{1g} (P) and T_{1g} (F) symbols are used. In the case of d^8 in octahedral field (under hole formulism), again ^3P is not split but ^3F is split. The order of energy level (due to splitting of ^3F state) of d^n and d^{10-n} ($n = 8$) have splitting pattern opposite to each other. So, A_{2g} is more stable than T_{2g} and T_{1g}.

On increasing ligand field strength, the gap between split states increases such that there is crossover between A_{2g} and T_{1g} (P) levels at high ligand field strength in d^2 octahedral case. But in d^8 octahedral case, due to interelectronic repulsion, bending of lines of energy level omits the crossing.

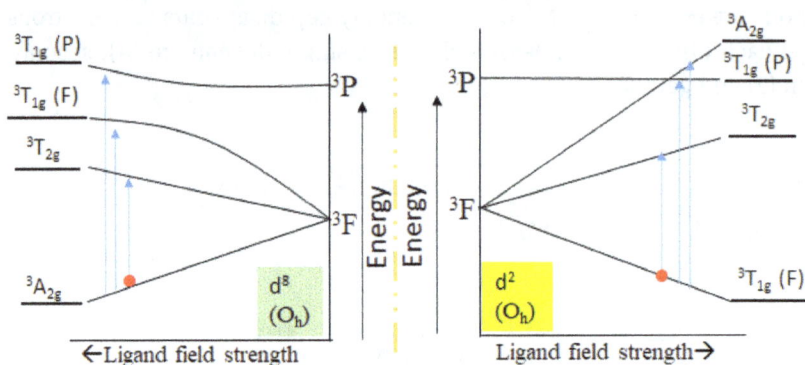

So, there should be three possible transitions in both unsymmetrical field but due to crossing of lines in octahedral field, only two peaks are found. Last two peaks $^3T_{1g}$ (F) to $^3T_{1g}$ (P) and $^3T_{1g}$ to $^3A_{2g}$ are very close to each other and so these are inseparable.

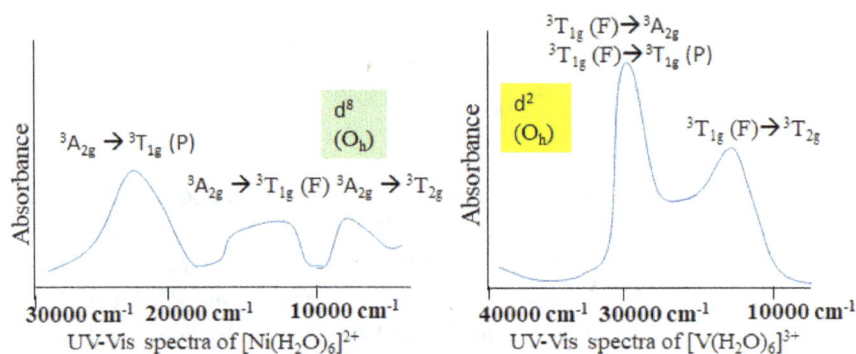

(f) Generalization of Orgel energy diagram

(i) Generalization based on d^1 system

Draw splitting pattern of any known d^n system, that is, d^1 in octahedral system. d^n and $d^{(5+n)}$ electrons have similar angular momentum. So, their splitting pattern are same, that is, d^1 and d^6 in octahedral system have similar splitting. d^n vis-à-vis $d^{(10-n)}$ are electron and equivalent hole set. Their splitting patterns are opposite of each other, that is d^1 vis-a-vis d^9 and d^6 vis-à-vis d^4 splitting patterns are opposite of each other. Splitting pattern of d^n tetrahedral is opposite of d^n octahedral or vice versa, that is, d^1 (t_d) has opposite splitting pattern than d^1 (O_h) and so on.

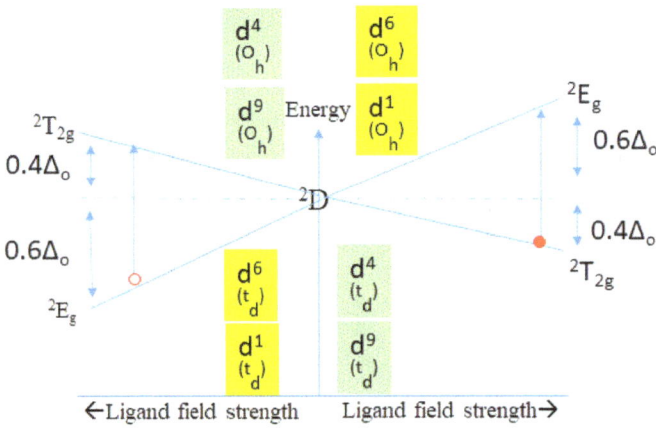

(ii) Generalization based on d^2 system

Draw splitting pattern of any known d^n system, that is d^2 in octahedral system. d^n and $d^{(5+n)}$ electrons have similar angular momentum. So, their splitting pattern is same, that is, d^2 and d^7 in octahedral system has similar splitting. d^n vis-à-vis $d^{(10-n)}$ are electron and equivalent hole set. Their splitting patterns are opposite of each other, that is, d^2 vis-à-vis d^8 and d^7 vis-à-vis d^3 splitting patterns are opposite of each other.

Splitting pattern of d^n tetrahedral is opposite of d^n octahedral or vice versa, that is, d^2 (t_d) has opposite splitting pattern than d^2 (O_h).

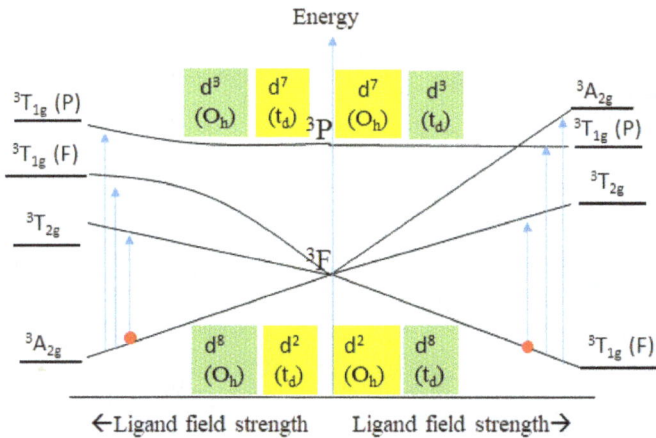

Annexure III

Case 3: d^5 complex system

In d^5 system, five electrons are rotating about their axes. There are different spin states of electrons in which maximum spin can be shown as following:

$$\boxed{\uparrow}\;\boxed{\uparrow}\;\boxed{\uparrow}\;\boxed{\uparrow}\;\boxed{\uparrow}$$

Maximum spin = $\Sigma s = 2\frac{1}{2}$
Different spin states = Σs, $(\Sigma s - 1)$ .... upto $0 = 2\frac{1}{2}, 1\frac{1}{2}, \frac{1}{2}$
Multiplicity (M): $2s + 1 = 6, 4, 2$

In d^5 system, five electrons are revolving around the nucleus. There are different angular states in which maximum angular state can be shown as:

$$\boxed{\uparrow\downarrow}\;\boxed{\uparrow\downarrow}\;\boxed{\uparrow}\;\boxed{}\;\boxed{}$$
$$m_l = +2,\ +1,\ 0,\ -1,\ -2$$

Maximum angular momentum = $\Sigma m_l = 6$
Different angular momentum states = Σm_l, $\Sigma m_l - 1$, upto $0 = 6, 5, 4, 3, 2, 1, 0$
Different angular momentum states $(L) = G, F, D, P, S$
Overall different total momentum states/energy states can be written as $^M L$.
Different energy states = ^6I, ^4I, ^2I, ^6H, ^4H, ^2H, ^6G, ^4G, ^2G, ^6F, ^4F, ^2F, $^6D, ^4D, ^2D$, ^6P, ^4P, ^2P, ^6S, ^4S, ^2S.

For multiplicity 6, we have to place one electron in each d component. That means it has angular momentum state 6.

$$\boxed{\uparrow}\;\boxed{\uparrow}\;\boxed{\uparrow}\;\boxed{\uparrow}\;\boxed{\uparrow}$$
$$m_l = +2,\quad +1,\quad 0,\ -1,\quad -2$$

Total spin (s): $\Sigma s = 2\frac{1}{2}$
Multiplicity (M): $2s + 1 = 6$
Angular states $(L) = \Sigma m_l = 0$
Energy states = $^{2S+1}L = {}^6S$

https://doi.org/10.1515/9783110656480-013

This spin state is only possible with S angular state (^6S).

For H and I, no other multiplicity than 2 is possible.

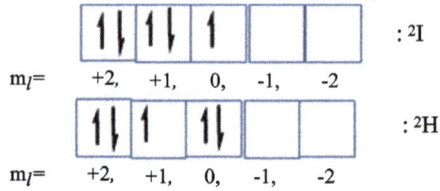

| $\uparrow\downarrow$ | $\uparrow\downarrow$ | $\uparrow$ | | | : ^2I |

m_l=   +2,   +1,   0,   -1,   -2

| $\uparrow\downarrow$ | $\uparrow$ | $\uparrow\downarrow$ | | | : ^2H |

m_l=   +2,   +1,   0,   -1,   -2

By any combination (same spinning electrons cannot be placed in same d component), ^4S will not be possible. So, the possible energy state will be ^2I, ^2H, ^4G, ^2G, ^4F, ^2F, ^4D, ^2D, ^4P, ^2P, ^6S, ^2S. Now, high spin states are more stable as 6 than 4 and 2 spin state. ^6S is the most stable. ^4G, ^4F, ^4D and ^4P, are moderately stable. ^2I, ^2H, ^2G, ^2F, ^2D, ^2P and ^2S are least stable.

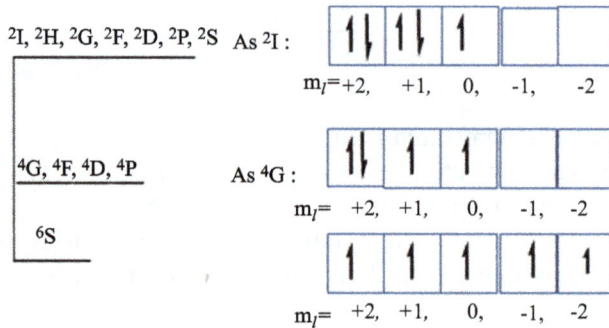

^2I, ^2H, ^2G, ^2F, ^2D, ^2P, ^2S As ^2I :

| $\uparrow\downarrow$ | $\uparrow\downarrow$ | $\uparrow$ | | |

m_l=+2,   +1,   0,   -1,   -2

^4G, ^4F, ^4D, ^4P As ^4G :

| $\uparrow\downarrow$ | $\uparrow$ | $\uparrow$ | | |

m_l=   +2, +1,   0,   -1, -2

^6S

| $\uparrow$ | $\uparrow$ | $\uparrow$ | $\uparrow$ | $\uparrow$ |

m_l=   +2, +1,   0,   -1, -2

Now, under spin selection rule, electrons favors transition without changing spin. But here, in any transition from ground state, reversal of spin takes place. So, all transitions are spin forbidden. Due to spin-forbidden transition, absorption bands are extremely weak. Transition between 6→4 multiplicity causes one electron spin reversal. So, absorption bands are weak but observable. Transition between 6→2 multiplicity causes double electron spin reversal. So, it is double spin forbidden and absorption bands are extremely weak (not observable). So, possible energy states for transition will be:

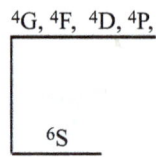

^4G, ^4F, ^4D, ^4P,

^6S

In same spin case, maximum exchange energy and minimum repulsion accounts the stability of energy states as:

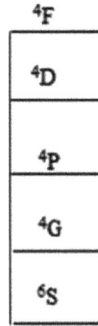

^4F

^4D

^4P

^4G

^6S

Spectra

The ground state ^6S does not split and transforms into $^6A_{1g}$. It is drawn along horizontal axis. ^4G, ^4F, ^4D and ^4P energy states constitute the excited energy states in the order of stability as $^4G > {^4P} > {^4D} > {^4F}$. Now, in the presence of asymmetric field (tetrahedral/octahedral), excited states are spitted as shown below (discussed earlier). ^4E_g, $^4A^{1g}$ (of ^4G), ^4E_g (of ^4D) and $^4A_{2g}$ (for 4F) energy levels are horizontal lines on the Orgel diagram. It indicates that these states are independent of the crystal field. Transition of electron from ground state to these excited states causes sharp peaks. Rest energy states are affected by crystal field. Increasing inclination of different energy states indicate that these states are more affected by crystal field. Transition of electron from ground state to these excited states causes broadening of peaks. Degree of broadening of peak is directly related to slope of energy state.

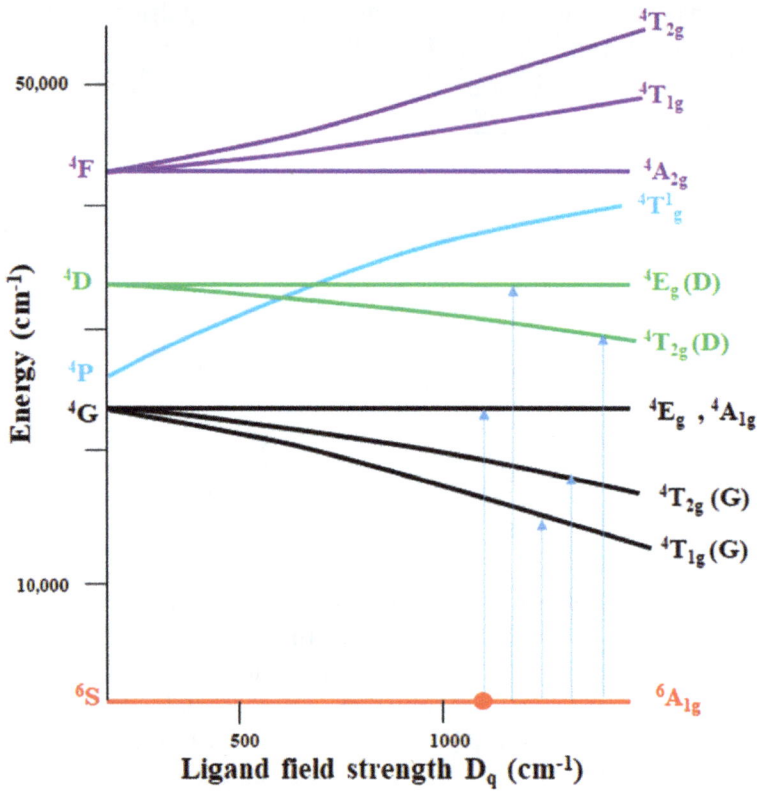

Energy state	Energy state splitting in unsymmetric field
D	E_g, T_{2g}
F	A_{2g}, T_{2g}, T_{1g}
G	A_{1g}, E_g, T_{1g}, T_{2g}
H	E_g, T_{1g}, T_{1g}, T_{2g}
I	A_{1g}, A_{2g}, E_g, T_{1g}, T_{2g}

Annexure IV

Bending of lines and calculation of Racah parameter and inseparable UV–vis band

One thing is noticeable in Orgel diagram of d^3 (O_h), d^7 (t_d), d^2 (t_d) and d^8 (O_h) system. $^3T_{1g}$ (F) and $^3T_{1g}(P)$ are very close in energy. Both have same symmetry and hence both undergo interelectronic repulsion pronouncedly. Interelectronic repulsion lowers the energy of lower state $^3T_{1g}$ (F) and increases the energy of higher state $^3T_{1g}(P)$.

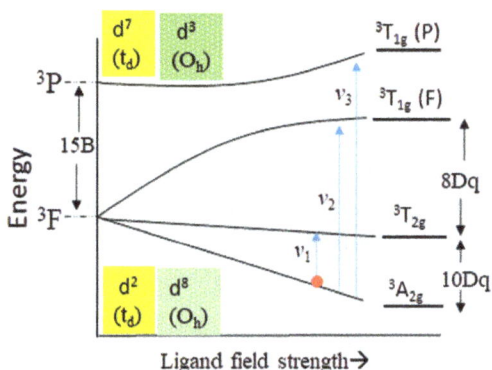

v_1, v_2, v_3 can be found by UV–vis spectra of $[CrF_6]^{3-}$.

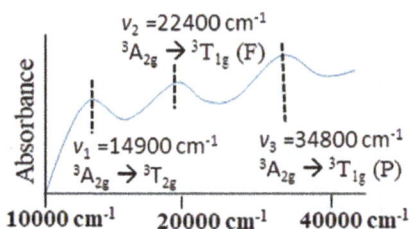

Let us take a case of d^3 (O^h): as $[CrF_6]^{3-}$. In the case of Cr^{3+} (d^3) case, the separation between ^3F and ^3P energy state is 15B. The energy gap (separation) between split terms of ^3F are 10Dq (for $^3A_{2g}$ and $^3T_{2g}$) and 8Dq (for $^3T_{2g}$ and $^3T_{1g}$). So, frequency of absorption should be as shown below:

$$v_1 = 10Dq$$

$$v_2 = 10Dq + 8Dq = 18Dq$$

$$v_3 = 10Dq + 2Dq + 15B$$

https://doi.org/10.1515/9783110656480-014

$^3T_{1g}$ (F) and $^3T_{1g}(P)$ undergo interelectronic repulsion and so, energy of upper state $^3T_{1g}(P)$ increases by x amount whereas energy of lower state $^3T_{1g}$ (F) decreases by x amount.

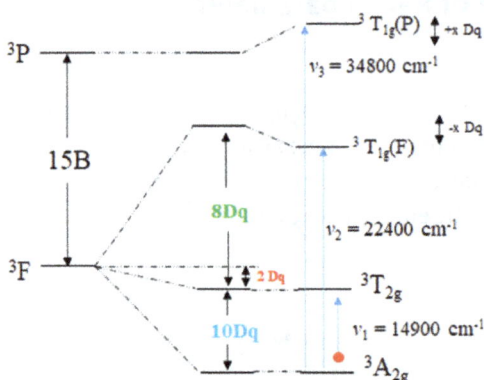

Now, energy gap between ^3F and ^3P states is shown by 15B', in which B' is apparent Racah parameter of complex ion. After including bending of split line, frequencies can be modified. It is actual frequency which is observed by UV–vis also.

$$v_1 = 10Dq = 14,900\,\text{cm}^{-1}$$

$$v_2 = 18Dq - x = 22,400\,\text{cm}^{-1}$$

$$v_3 = 12Dq + 15B' + x = 34,800\,\text{cm}^{-1}$$

In the term, of B' and Dq, the above equations are written as

$$v_1 = 10Dq \tag{1}$$

$$v_2 = 7.5B' + 15Dq - 0.5(225B'^2 + 100Dq^2 - 180B'Dq)^{0.5} \tag{2}$$

$$v_3 = 7.5B' + 15Dq + 0.5(225B^2 + 100Dq^2 - 180B'Dq)^{0.5} \tag{3}$$

By solving first two equations (1 and 2), value of B' and Dq can be found.

If third peak is not separable in UV band, then it can be approximated by putting the value of B' and Dq in eq. (3). Overall, apparent Racah parameter B' is the expression for interelectronic repulsion in metal complex ion whereas Racah parameter B is the expression for interelectronic repulsion in metal ion. Ratio of both B' and B is called nephelauxetic ratio (β)

$$\beta = B'/B$$

(Value of B is already indexed in texts).

Annexure V

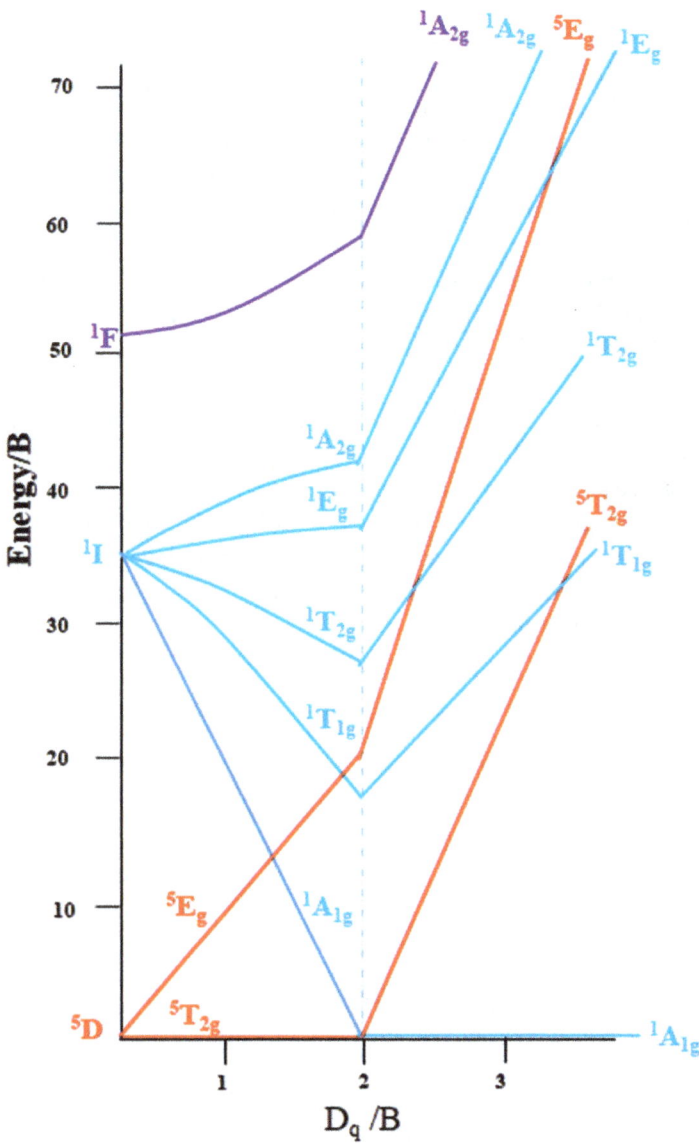

Tanabe–Sugano diagrams for d^6 complex

Tanabe–Sugano diagrams are used for both weak field as well as strong field case. Here, we have considered d^6 octahedral high spin complex $[CoF_6]^{3-}$ (where F^- provides weak ligand field) and d^6 octahedral low spin complex $[Co(en)_3]^{3+}$ (where

https://doi.org/10.1515/9783110656480-015

ethylenediamine provides strong ligand field). There are different energy states in Co^{3+} ionic states, that is d^6 octahedral complex (Co^{3+}) has ^5D (ground states) and ^1I, ^1F excited energy states. In the presence of unsymmetrical field, energy states split into several energy levels. ^5D ground state splits into ^5E_g and $^5T_{2g}$ energy levels. ^1I splits into five energy levels A_{1g}, A_{2g}, E_g, T_{1g} and T_{2g}.

On increasing ligand field strength, the gap between energy levels increases. Some energy levels get up in energy whereas some get down in energy. ^5D state: ^5E_g energy level gets up in energy whereas $^5T_{2g}$ energy level remains horizontal. ^1I state: $\boldsymbol{A_{2g}}$ and $\boldsymbol{E_g}$ energy levels get up in energy whereas A_{1g} T_{1g} and T_{2g} energy levels get down in energy.

Among all energy level, one energy level shows strong dependence with crystal field. With increasing crystal field strength (Dq/B), this particular energy level falls rapidly and reaches a null value of E/B (energy per unit field). This point indicates spin pairing of electrons.

^1I state: A_{1g} falls rapidly with ligand field. In other words, it can be said that A_{1g} level is greatly stabilized by ligand at Dq/B = 2. A_{1g} energy level gets E/B = 0 at Dq/B = 2. In other work, it can be said that after crystal field strength Dq/B = 2, electrons are paired up and it becomes ground level for strong filed ligands (low spin complex). On the expanse of this paring in A_{1g} energy level, rest energy levels shoot up in energy. The energy profile is redrawn for such low-spin complexes. This point is called high spin–low spin crossover point.

Annexure VI

Tanabe–Sugano diagrams: theory and guess of B and Dq

The Orgel energy-level diagram is used for weak field case whereas Tanabe–Sugano diagrams are used for both weak field as well as strong field case. The Orgel energy-level diagram is only useful for spin-allowed transition, whereas the Tanabe–Sugano diagrams are useful for other transitions also. In the Tanabe–Sugano diagram, ground state term is taken as horizontal axis. This horizontal line provides constant reference point and the other states can be plotted relative to ground state. The Orgel energy-level diagram is plot of energy of state (E) versus strength of field (Dq), whereas the Tanabe–Sugano diagram is the plot of E/B versus Dq/B. Let us start with an example: The Orgel diagram for $[Ni(H_2O)_6]^{2+}$ d^8 octahedral system can be shown as:

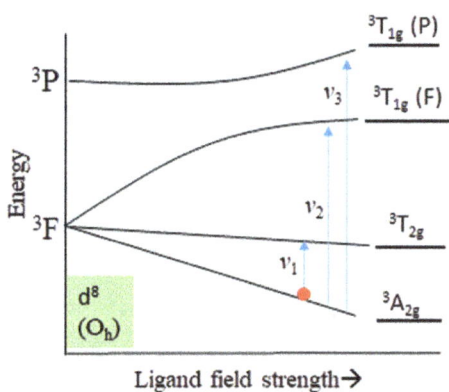

Peak of UV–vis spectra of $[Ni(H_2O)_6]^{2+}$ can be correlated according to the Orgel diagram energy gap.

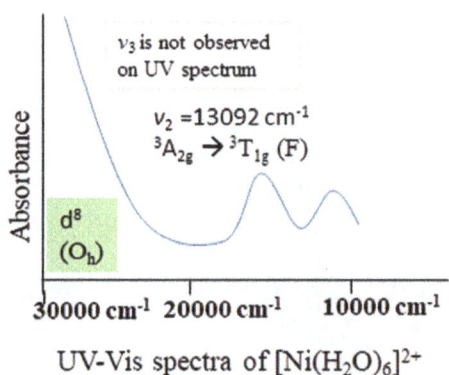

UV-Vis spectra of $[Ni(H_2O)_6]^{2+}$

https://doi.org/10.1515/9783110656480-016

The ratio of v_2/v_1 can be calculated from UV–vis spectra as follows:

$$v_2/v_1 = E_2/E_1 = 13{,}092/8{,}143 = 1.608$$

It is equal to $(E_2/B)/(E_1/B) = 1.608$

As first guess, take the scale at $Dq/B = 1$. Then, E/B can be found on vertical axis

$$E_1/B = 10.1$$

$$E_2/B = 16.6$$

$(E_2/B)/(E_1/B) = 1.64$ It is somewhat high then required. Now, as second guess, take the scale at $Dq/B = 1.5$. Then, E/B can be found on vertical axis

$$E_1/B = 14.9$$

$$E_2/B = 22.3$$

$$(E_2/B)/(E_1/B) = 1.50$$

It is less than required. That means the exact guess will be more toward 1 and less toward 1.5 Dq/B.

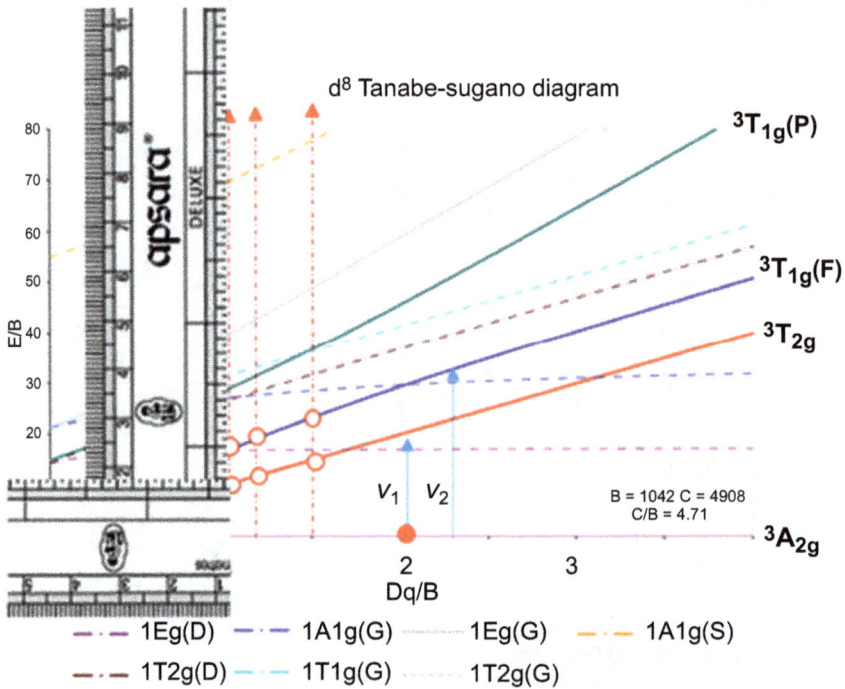

d^8 Tanabe-sugano diagram

Now, make a guess in between the previous two guesses, $Dq/B = 1.25$. Then, E/B can be found on vertical axis

$$E_1/B = 12.6$$

$$E_2/B = 20.2$$

$$(E_2/B)/(E_1/B) = 1.60$$

This one is a good guess at $Dq/B = 1.25$.

Now, Racah B parameter can be calculated easily;

$$E_1/B = 8,143/B = 12.6$$

$$B = 646 \, \text{cm}^{-1}$$

$$E_2/B = 13,092/B = 12.6$$

$$B = 648 \, \text{cm}^{-1}$$

Now, value of Dq can be calculated as follows:

$$Dq/B = 1.25; \, Dq = 1.25 \times B = 1.25 \times \, 647 \, \text{cm}^{-1} = 8,090 \, \text{cm}^{-1}$$

Index

https://doi.org/10.1515/9783110656480-017